ENDANGERED
OCEANS

Opposing Viewpoints®

OTHER BOOKS OF RELATED INTEREST

OPPOSING VIEWPOINTS SERIES

Endangered Species
The Environment
Global Resources
Global Warming

CURRENT CONTROVERSIES SERIES

Conserving the Environment
Pollution

ENDANGERED OCEANS

Opposing Viewpoints®

William Dudley, Book Editor

David L. Bender, Publisher
Bruno Leone, Executive Editor
Bonnie Szumski, Editorial Director
David M. Haugen, Managing Editor

OPPOSING
VIEWPOINTS®
SERIES

Greenhaven Press, Inc., San Diego, California

Cover photo: Digital Stock

Library of Congress Cataloging-in-Publication Data

Endangered oceans : opposing viewpoints / William Dudley, book editor.
    p.   cm. — (Opposing viewpoints series)
  Includes bibliographical references and index.
  ISBN 0-7377-0063-7 (lib.). — ISBN 0-7377-0062-9 (pbk.)
  1. Marine ecology. 2. Endangered ecosystems. 3. Marine resources
conservation. 4. Fishery conservation. 5. Whales. 6. Endangered
species. I. Dudley, William, 1964 –   . II. Series: Opposing viewpoints
series (Unnumbered)
QH541.5.S3E535 1999
577.7—dc21   98-45933
                                                                  CIP

Greenhaven Press, Inc., P.O. Box 289009
San Diego, CA 92198-9009

"CONGRESS SHALL MAKE NO LAW...ABRIDGING THE FREEDOM OF SPEECH, OR OF THE PRESS."

First Amendment to the U.S. Constitution

The basic foundation of our democracy is the First Amendment guarantee of freedom of expression. The Opposing Viewpoints Series is dedicated to the concept of this basic freedom and the idea that it is more important to practice it than to enshrine it.

CONTENTS

**Chapter 3: What Policies Would Best Protect the
World's Fisheries?**

Chapter 4: How Can Whales Best Be Protected?

WHY CONSIDER
OPPOSING VIEWPOINTS?

"*The only way in which a human being can make some
approach to knowing the whole of a subject is by hearing
what can be said about it by persons of every variety of
opinion and studying all modes in which it can be looked
at by every character of mind. No wise man ever acquired
his wisdom in any mode but this.*"

John Stuart Mill ·

In our media-intensive culture it is not difficult to find differing
opinions. Thousands of newspapers and magazines and dozens
of radio and television talk shows resound with differing points
of view. The difficulty lies in deciding which opinion to agree
with and which "experts" seem the most credible. The more inundated we become with differing opinions and claims, the
more essential it is to hone critical reading and thinking skills to
evaluate these ideas. Opposing Viewpoints books address this
problem directly by presenting stimulating debates that can be
used to enhance and teach these skills. The varied opinions contained in each book examine many different aspects of a single
issue. While examining these conveniently edited opposing
views, readers can develop critical thinking skills such as the
ability to compare and contrast authors' credibility, facts, argu-
mentation styles, use of persuasive techniques, and other stylis-
tic tools. In short, the Opposing Viewpoints Series is an ideal
way to attain the higher-level thinking and reading skills so essential in a culture of diverse and contradictory opinions.

In addition to providing a tool for critical thinking, Opposing
Viewpoints books challenge readers to question their own
strongly held opinions and assumptions. Most people form their
opinions on the basis of upbringing, peer pressure, and personal, cultural, or professional bias. By reading carefully balanced opposing views, readers must directly confront new ideas
as well as the opinions of those with whom they disagree. This
is not to simplistically argue that everyone who reads opposing
views will—or should—change his or her opinion. Instead, the
series enhances readers' understanding of their own views by
encouraging confrontation with opposing ideas. Careful examination of others' views can lead to the readers' understanding of
the logical inconsistencies in their own opinions, perspective on

why they hold an opinion, and the consideration of the possibility that their opinion requires further evaluation.

EVALUATING OTHER OPINIONS

To ensure that this type of examination occurs, Opposing Viewpoints books present all types of opinions. Prominent spokespeople on different sides of each issue as well as well-known professionals from many disciplines challenge the reader. An additional goal of the series is to provide a forum for other, less known, or even unpopular viewpoints. The opinion of an ordinary person who has had to make the decision to cut off life support from a terminally ill relative, for example, may be just as valuable and provide just as much insight as a medical ethicist's professional opinion. The editors have two additional purposes in including these less known views. One, the editors encourage readers to respect others' opinions—even when not enhanced by professional credibility. It is only by reading or listening to and objectively evaluating others' ideas that one can determine whether they are worthy of consideration. Two, the inclusion of such viewpoints encourages the important critical thinking skill of objectively evaluating an author's credentials and bias. This evaluation will illuminate an author's reasons for taking a particular stance on an issue and will aid in readers' evaluation of the author's ideas.

As series editors of the Opposing Viewpoints Series, it is our hope that these books will give readers a deeper understanding of the issues debated and an appreciation of the complexity of even seemingly simple issues when good and honest people disagree. This awareness is particularly important in a democratic society such as ours in which people enter into public debate to determine the common good. Those with whom one disagrees should not be regarded as enemies but rather as people whose views deserve careful examination and may shed light on one's own.

Thomas Jefferson once said that "difference of opinion leads to inquiry, and inquiry to truth." Jefferson, a broadly educated man, argued that "if a nation expects to be ignorant and free . . . it expects what never was and never will be." As individuals and as a nation, it is imperative that we consider the opinions of others and examine them with skill and discernment. The Opposing Viewpoints Series is intended to help readers achieve this goal.

David L. Bender & Bruno Leone,
Series Editors

Greenhaven Press anthologies primarily consist of previously published material taken from a variety of sources, including periodicals, books, scholarly journals, newspapers, government documents, and position papers from private and public organizations. These original sources are often edited for length and to ensure their accessibility for a young adult audience. The anthology editors also change the original titles of these works in order to clearly present the main thesis of each viewpoint and to explicitly indicate the opinion presented in the viewpoint. These alterations are made in consideration of both the reading and comprehension levels of a young adult audience. Every effort is made to ensure that Greenhaven Press accurately reflects the original intent of the authors included in this anthology.

INTRODUCTION

"We have the power to undermine the healthy functioning of the sea that supports all life on Earth."

Sylvia Earle, marine biologist and former chief scientist
of the National Oceanic and Atmospheric Administration

The great oceans of the world—the Atlantic, Pacific, Indian, and Arctic Oceans—are in reality one connected body of water that covers almost 71 percent of the world's surface. This global ocean, twelve thousand feet deep on average, holds almost 98 percent of the planet's water and is home to an estimated 80 percent of the world's biodiversity.

The oceans provide humanity with numerous benefits. They have served as important transportation and trade highways since the invention of ships thousands of years ago. Oceans have been historically utilized as convenient garbage dumps for waste ranging from municipal sewage to surplus military weapons. Humans have also exploited the oceans for resources ranging from petroleum to seafood. Offshore wells provide about 17 percent of the world's oil and natural gas production. Humans gather ninety million tons of fish and shellfish annually for food. Kelp, an ocean plant, is used for food and for algin, a thickening substance found in many food and cosmetic products. Ocean plants and animals are also important sources of medicines and anticancer treatments.

In addition to these resources, the global ocean serves several important ecological functions that help make life on this planet possible—functions that are only beginning to be fully understood by scientists. Its enormous volume of water, which circulates in large horizontal and vertical currents, helps to stabilize temperatures in the Earth's atmosphere. Evaporated ocean water is the source of most of the world's precipitation. In addition, the oceans, through the actions of floating microscopic plant life (plankton), serve as the planet's "lungs" by absorbing carbon dioxide and releasing oxygen. This not only provides much of the oxygen necessary for all animal life, but also alleviates carbon dioxide's "greenhouse effect" and prevents global warming.

Many people until recently have assumed that the oceans and their resources were so large as to be impervious to human actions. The global ocean was considered to be "vast, immutable, an endless storehouse of resources and a bottomless receptacle for our wastes," according to environmental writer Kieran Mulvaney. "But for all its dimensions," Mulvaney continues, "the ocean is not immune to the effects of human activities." Threats to the global ocean's environment, in the view of Mulvaney and other concerned observers, include the taking of too many fish for natural replenishment (overfishing), the befouling of ocean waters with oil, toxic chemicals, garbage, and other pollutants, development on land (including buildings and dams) that disrupts wetlands, beaches, and other coastal ecosystems, the introduction of exotic species to marine environments, and global climate change that can affect sea levels and ocean currents.

While most people agree that human activities such as overfishing and pollution can at least potentially endanger the ecological balance of the world's oceans, profound disputes remain over what should be done to curb these threats. Some argue that protecting the oceans is a political challenge to the global community that must be met through increased national regulations and international treaties. For others, the problems—and solutions—concerning the oceans are fundamentally economic and relate to the lack of private property rights on the seas.

Sam Farr, a member of Congress representing California, argues that oceans are endangered because users of marine resources act free of supervision or regulation. "It's the last frontier where you can do whatever you want, and that's what is getting us into trouble." Such laws that do exist are often too loosely enforced by a complex patchwork of governments, Farr and others contend. In the United States, for example, local governments regulate activities close to the shore, state governments have jurisdiction extending out twelve miles, and the federal government controls activities two hundred miles from land. "The choreography needs to be improved," according to Farr, who has called for a comprehensive review of U.S. oceans policy. The Independent World Commission on the Oceans, an advisory body of experts and political leaders, goes further to argue that marine issues transcend national boundaries and require greater involvement in ocean governance by the United Nations and other international institutions. Among the commission's recommendations is the creation of a World Ocean Affairs Observatory to monitor activities on the seas and help ensure that international laws on fishing, pollution, and other matters are being obeyed. Mario Soares, chairperson of the commission, argued that "we need to ensure that management of the oceans is rational, just, and responsible vis-à-vis future generations." This requires, in his view "an integrated and consensual governance of the oceans under the aegis of the United Nations."

Some observers, however, argue that problems created by human activities on the oceans stem not from lack of external regulation, but from perverse economic incentives. Entrepreneur Michael Markels Jr. argues that a significant obstacle to sustainable ocean development is "the lack of property rights in the seas." Because "single individuals and firms do not have exclusive claim to portions of the sea's bounty," he asserts, "there is every incentive for any one fisherman to catch as many fish as possible. After all, if he does not, someone else will. If there is one fish left, it is to the individual fisherman's advantage to catch it." Environmental analyst Kent Jeffreys agrees, contending that the oceans provide a classic example of the "tragedy of the commons" in which a resource is under common ownership. "If everyone has a right to exploit a resource," he argues, "it will rapidly be degraded by overuse. If no one has the responsibility for maintaining the resource, no one will do so." The solution, Markels, Jeffreys, and others suggest, is to introduce market forces and privatize parts of the ocean in a way similar to how much of the world's land surface is divided up into private property. "A property rights approach," claims Jeffreys, "can empower individuals to protect and utilize fishery resources simultaneously. . . . It is time to determine who should own the ocean."

Not everyone agrees that private property rights and market forces can best protect the world's oceans. According to the Independent World Commission on the Oceans, "global market forces—while generating an extraordinary energy for growth—have often substituted themselves for government with some unfortunate results. These forces tend to be insensitive to the social and ecological harm that is being caused, in particular to the ocean." The oceans, the commission concludes, must "be regarded as a common resource so as to make it possible for nations and people to share more equitably in the benefits of resource exploitation."

Disagreements over the sharing and managing of the oceans' resources are one of several points of contention raised in the following chapters of *Endangered Oceans: Opposing Viewpoints*: How Endangered Are the World's Oceans and Coastlines? What Ocean Management and Conservation Practices Are Best? What Policies Would Best Protect the World's Fisheries? How Can Whales Best Be Protected? The contributors to this volume agree on the importance of oceans to both the world economy and environment, but provide diverse opinions as to how the global ocean and its resources can best be protected.

HOW ENDANGERED ARE THE WORLD'S OCEANS AND COASTLINES?

CHAPTER PREFACE

The areas where the oceans and the continents meet are important for both humans and ocean wildlife. About two-thirds of the world's people live less than 150 kilometers from the ocean. The waters off the world's coastlines, with their estuaries, lagoons, coral reefs, and mangrove forests, are home to many of the ocean's most biologically rich environments and provide much of the seafood humans consume. Many ocean species spend at least part of their life cycles in coastal zones.

Unfortunately, the coasts are also the part of the ocean that is most heavily polluted by humans. Many of the toxic chemicals used in the nation's farms, factories, and homes eventually reach the ocean, accounting for three-quarters of pollution in coastal waters. The Environmental Protection Agency estimates that in the United States alone 32 billion gallons of runoff—contaminated water drained from farms, factories, and urban areas—enter the oceans each day. Agricultural runoff often contains manure, soil, and fertilizers that can stimulate the growth of dense algae blooms. These blankets of algae harm other aquatic life by cutting off sunlight and depleting ocean water's oxygen content. Urban runoff can include discarded motor oil, garbage washed down storm drains, and other refuse that can poison sea life and make ocean water unsafe for humans to swim in.

Cleanup measures have been at least partially successful in alleviating the problem of water pollution. Countries have agreed in international treaties to regulate ocean dumping and ban the disposal of radioactive materials in the ocean. The United States, for example, stopped all ocean dumping of sewage and garbage by 1992. Water quality has improved in some areas of the United States. Since 1986, scientists participating in the Mussel Watch Project of the National Oceanic and Atmospheric Administration have periodically analyzed the concentrations of pollutants absorbed by mussels and oysters that inhabit coastal waters; their studies have shown decreasing levels of cadmium, polychlorinated biphenyls (PCBs), and other metals and chemicals.

The oceans constitute an important part of the earth's environment. The extent to which the world's oceans, including the vital coastal areas, are ecologically endangered is examined in the viewpoints of this chapter.

| "Vital coastal habitats are being buried, damaged, altered or destroyed by construction and development."

THE FUTURE OF THE WORLD'S OCEANS AND COASTAL AREAS IS BLEAK

Kieran Mulvaney

Kieran Mulvaney is a freelance writer who has written numerous articles on marine and environmental issues. In the following viewpoint, he writes that people once believed that the world's oceans were so vast as to be unaffected by human activity. However, he asserts, over the past few decades overfishing, air and water pollution, and coastal development have inflicted much ecological damage on the world's oceans and coastlines. Mulvaney furthermore contends that national and international regulations have done little to stop the environmental degradation of the oceans.

As you read, consider the following questions:

1. How large is the global ocean in relation to the rest of the earth, according to scientist James Lovelock, as reported by the author?
2. What various forms of ocean pollution does Mulvaney describe?
3. Why have some environmental organizations been disappointed with political solutions to ocean problems, according to Mulvaney?

Abridged from Kieran Mulvaney, "A Sea of Troubles: In the International Year of the Ocean, Are We Reaching Our Limits?" E/The Environmental Magazine, January/February 1998. Reprinted with permission from E/The Environmental Magazine; Subscription Department: PO Box 2047, Marion, OH 43306; telephone: (815) 734-1242. Subscriptions are $20 per year.

Ours is a water planet. The ocean covers 71 percent of the surface area of the globe, and constitutes over 90 percent of all habitable space on Earth. Its total volume is around 300 million cubic miles and its weight is approximately 1.3 million million million tons. No wonder that Arthur C. Clarke, scientist and writer, once remarked that it was "inappropriate to call this planet Earth, when clearly it is ocean."

THE GLOBAL OCEAN

The vast dimensions of the global ocean moved one scientist to suggest 40 years ago that it "may be rash to put any limit on the mischief of which man is capable, but it would seem that those 100 and more million cubic miles of water . . . is the great matrix that man can hardly sully and cannot appreciably despoil."

But those "100 and more million cubic miles" need to be put into perspective. As Jim Lovelock, originator of the Gaia hypothesis, has observed, "Although the weight of the oceans is 250 times that of the atmosphere, it is only one part in 4,000 of the weight of the Earth." If the Earth were a globe 12 inches in diameter, notes Lovelock, the average depth of the ocean would be no more than the thickness of a piece of paper, and even the deepest ocean trench would be a dent of a third of a millimeter.

Even so, it is easy to understand the reasoning behind the logic of that 1950s scientist, Dr. Sylvia Earle, former chief scientist for the National Oceanic and Atmospheric Administration (NOAA), points out that "As recently as a half century ago, the sea still seemed to be in excellent health physically, chemically and biologically. When the explorer Thor Heyerdahl sailed in 1947 with a crew of five others across the Pacific Ocean from Peru to Tahiti, weeks passed with no clues to suggest that humankind existed anywhere except on their raft."

But, says Earle, by 1970, when Heyerdahl set out on another raft journey, this time across the Atlantic, something of a "sea change" was already underway. "He reported seeing far more oil lumps than fish, and alerted the world about the enormous quantities of trash, oily wastes and plastic debris he observed in the sea."

Heyerdahl was a harbinger of deepening bad news for the world's oceans. Since the 1970s, commercial fisheries have pushed fish stocks to collapse. Pollution has claimed the lives of millions of seabirds, and untold numbers of birds, marine mammals and sea turtles become entangled or ensnared each year in plastic debris that finds it way into the sea. Vital coastal habitats are being buried, damaged, altered or destroyed by construction and development.

In response, the United Nations has declared 1998 the International Year of the Ocean. Across the globe, scientists, environmentalists and others are training their focus on the array of human impacts that are making themselves felt on the global ocean.

FISHERIES: REACHING DEPLETION

According to the United Nations Food and Agriculture Organization (FAO), an estimated 70 percent of global fish stocks are "over-exploited," "fully exploited," "depleted" or recovering from prior over-exploitation. By 1992, FAO had recorded 16 major fishery species whose global catch had declined by more than 50 percent over the previous three decades—and in half of these, the collapse had begun after 1974. In 1992, the virtual disappearance of Northwest Atlantic groundfish led the Canadian government to close commercial fisheries and, later, all fishing on these stocks. A 1997 paper in the British journal *Nature* predicted that, unless swift and effective action was taken to protect them, cod stocks in the North Sea were also in danger of collapse. At least one species—the California white abalone—is now considered a likely candidate for extinction, 20 years after intense exploitation ended.

At the same time, as much as 27 million tons of fish are thrown overboard annually because they are undersized, of the wrong species, of inferior quality or surplus to quotas. A study in Alaska suggests that Bering Sea red king crab discards amounted to 16 million animals in 1990, more than five times the number actually landed.

Large numbers of marine mammals, sea turtles and seabirds are also caught in commercial fisheries operations around the world. The National Research Council has identified bycatch in shrimp trawls as the most significant cause of sea turtle mortality in the U.S. Tuna long-line fisheries in the Southern Ocean are estimated to entangle at least 44,000 albatrosses every year, and possibly many more. Harbor porpoises are caught in large numbers virtually everywhere gill nets are set in coastal waters.

Aquaculture, or fish farming, which is often touted as a panacea for the problems of fisheries over-exploitation, is not necessarily an answer. The construction of aquaculture facilities can result in the loss and fragmentation of habitats, particularly mangrove forests. Fish farms also often result in high levels of nutrient and chemical pollution and the escape of introduced fish species and associated diseases into the wild. In addition, large numbers of wild fish are caught to feed those raised in farms: for example, the production of one ton of cage-reared salmon requires approximately 5.3 tons of fish. The over-exploitation of stocks for fishmeal is considered the likely cause of the dramatic collapse of some seabird populations in the North Sea region during the 1980s.

POLLUTION: OUR GLOBAL GARBAGE CAN

Pollution of the ocean comes in many and varied forms, and from a wide range of sources. The National Research Council has estimated that as many as 8.8 million tons of oil enter the ocean each year as a result of human activity, and that at any given time, the ocean contains 280,000 tons of tar balls. All kinds of garbage, ranging from fishing nets to trash from cargo ships to litter on the beach, finds its way into coastal waters and the ocean, where it traps, ensnares and entangles marine wildlife such as marine mammals, sea turtles and seabirds. Plastic pellets have been found on the surface of the Pacific at concentrations of 21,000 per square mile; a clean-up exercise on the coast of Texas yielded 15,600 six-pack rings along 1.8 miles of coastline; and a National Academy of Sciences review once estimated that over 14 billion pounds of garbage enters the ocean from sea-based sources alone. In the 1980s, it was reckoned that 30,000 northern fur seals died each year after becoming entangled in marine debris, principally lost or abandoned fishing gear.

Heavy metals—for example, mercury and lead—and organo-chlorine compounds such as PCBs and DDT have been associated with a wide range of impacts on marine wildlife.

According to Boyce Thorne-Miller, senior scientist with Sea-Web, a marine conservation education initiative of The Pew Charitable Trusts, "Although it's difficult to definitively establish cause and effect in a lot of these cases, these contaminants have been linked with mortality, malformation, reduced hatching success, developmental abnormalities and chromosome aberrations in fish eggs and larvae contaminated at the surface, and reproductive problems and reduced immune system in marine mammals." Because heavy metals and organochlorines are bioaccumulative, that is, they build up in progressively greater concentrations as they are passed up the food chain. Top-line predators are particularly at risk, and their plight has been taken up by the new Ocean Wildlife campaign. Striped dolphins in the western North Pacific, for example, have concentrations of PCBs and DDT more than 10 million times higher than that of the water they live in.

COASTAL HABITAT DESTRUCTION: PUSHED BY POPULATION

The fate of the ocean is inextricably entwined with that of the coast. "The coasts," says Beth Milleman of the Washington,

D.C.-based Coast Alliance, "have been described as underwater rainforests because of the incredible diversity of life they contain, and there's a lot of truth to that."

Many ocean species rely on coastal habitats for breeding, feeding and shelter: one-third of the world's marine fish species are found on coral reefs, the most productive coastal ecosystems of all, and it has been estimated that the total number of species of all kinds in reef systems could number a million. Other coastal habitats, such as mangroves and sea grasses, are also vital breeding, feeding and nursery areas for fish and shellfish species, home to a variety of wildlife species, and important protection and shelter against storms and coastal erosion. Ninety percent of the current world fisheries harvest comes from within 200 miles of the coast, and most of that within a strip of just five miles from the coast.

But the coastal zone is also home to the majority of the world's population. As much as 66 percent of the world's population lives within 40 miles of the shore, and coastal populations are growing faster than the global population as a whole. In the U.S. between 1960 and 1990, the population in coastal counties grew by 41 million, an increase of 43 percent. Between 1983 and 1991, 90 percent of all building activity in Australia took place within the coastal zone.

As a result of such growth in population and development, among other factors, coastal environments are coming under increasing pressure. It is estimated, for example, that as much as 10 percent of the world's coral reefs have been degraded beyond recovery, and that another 30 percent is likely to decline within the next 15 or so years. Seventy-five percent of mangrove forests in the Philippines, and 40 percent in Ecuador, have been cut down to make way for aquaculture ponds. Around the world, seagrasses are being stifled by turbidity in the water as a result of nutrient pollution.

By interrupting the flow of freshwater from rivers, the construction of dams has impacted coastal regions and destroyed the habitats of many fish species worldwide: they are considered, for example, to be one of the primary causes in the extinction of at least 106 major populations of salmon and steelhead on the west coast.

THE HAVOC OF EXOTIC MIGRATION

Although still an obscure problem, the constant introduction of exotic species to marine environments where they do not naturally occur is, says Dr. James Carlton, professor of marine science at Williams College–Mystic Seaport, Connecticut, playing "ecological roulette with the ocean. There is no way of knowing where and when the next invasion will occur, or what the consequences will be. But we do know that every time we introduce a species, we run the risk of radically transforming marine ecosystems, with tremendous ecological, economic and social consequences."

The principal method by which exotic species are introduced into marine environments is through the intake and discharge of ballast water. When ships take on ballast at their point of departure, they also take on board thousands of microscopic organisms, including the planktonic life stages of larger plants and animals. As the ballast is emptied at the port of call, these passengers are discharged as well.

"We reckon that, at any time, there are 3,000 species in motion in ballast water," says Carlton, "and that, somewhere in the world, one introduced species is taking hold every day."

One dramatic example is the Atlantic comb jelly, a U.S. east coast native, introduced by ballast water into the Black and Azov Seas in the early 1980s. By 1988, it had become the dominant species in the Black Sea, leading to collapses in fish stocks and an estimated $250 million of lost fisheries revenue. Introduced species have also transformed marine ecosystems in the U.S.: there are at least 250 exotic organisms in San Francisco Bay alone, including the Asian clam, which is now found at densities of 3,000 per square foot.

The International Maritime Organization (IMO) is looking at ways to regulate ballast water discharge, and researchers in Australia and the United States are finding ways to tackle the problem by using heat to kill organisms in ballast water, or developing filters to trap the organisms when the ballast is discharged or taken on board. The island nation of Bonaire prohibits the dumping of ballast water in its coastal waters. But it is, admits Carlton, like "pushing a peanut uphill," and in the meantime, more catastrophic species introductions seem certain to occur.

CLIMATE CHANGE AND GLOBAL WARMING

Finally, all these separate threats need to be placed in the context of overall global change, with an altered climate and increased ultraviolet radiation as a result of ozone depletion two prime examples.

According to a review by the Intergovernmental Panel on Climate Change (IPCC), a grouping of some 300 scientists from around the world, climate change "has the potential to significantly affect biological diversity in ocean and coastal areas. It could cause changes in the population sizes and distributions of species, alter the species composition and geographical extent of habitats and ecosystems, and increase the rate of species extinctions."

These changes could come about, says the IPCC, through any combination of sea-level rise, increases in sea-surface temperature, increases in storms and other extreme events, and increased precipitation leading to greater run-off of pollutant-and-nutrient-rich soil and water into coastal areas. For example, rising sea-levels may swamp coastal habitats, and higher sea surface temperatures have already been implicated in some coral diseases and in nurturing some harmful algal blooms.

In addition, there is growing evidence that increased levels of ultraviolet-B (UV-B) radiation as a result of ozone depletion may be harming marine species, particularly those in the upper layers of the sea. Numerous studies have shown, for example, that increased UV-B can cause death, decreased reproductive capacity, reduced survival and impaired larval development in some of the plankton species that form the basis of the marine food chain.

The Healing Process

Given the size and extent of the ocean, and the complexity and variety of the issues it faces, addressing threats to the marine environment generally requires a multifaceted approach. Because of the global nature of human activities that impact the ocean, many environmentalists concentrate their efforts on seeking to have those activities regulated or, if necessary, banned by international conventions.

Unfortunately, observes Clifton Curtis, political advisor to Greenpeace International, "There remains a tendency on the part of international agreements to put the ocean in a box and say, 'OK, we've done rainforests, now let's address oceans.' But 'ocean issues' cover such a wide range—fisheries, oil and gas, minerals, to name a few—that you can't just fence them off that neatly."

That said, Curtis does see progress in the willingness of some countries to begin addressing those issues. Specifically, he cites the recent United Nations Convention on Straddling Fish Stocks and Highly Migratory Fish Stocks, developed to deal with the thorny issue of fisheries whose targets straddle or migrate between countries' national waters and the high seas; the entry into force of the UN Law of the Sea, which covers a huge array of subjects, from navigation rights to fisheries to seabed mining; the interest of established agreements, such as the Convention on Biological Diversity and the Commission on Sustainable Development, in supporting ocean conservation; and the development, under the leadership of the United Nations Environment

Program (UNEP), of a broad-based Global Program of Action for the Protection of the Marine Environment from Land-Based Activities.

Unfortunately, Curtis admits, it is often one thing for countries to adopt strict-sounding rules and regulations, and quite another to show the political will to enforce them. "For example, when it became clear that the Soviet Union had been dumping large amounts of radioactive material in the Kara and Barents Sea, in direct violation of the London Convention, very little was done. Certainly, no punitive measures were taken."

COASTS IN CRISIS

In the Philippines, subsistence fishermen dynamite irreplaceable coral reefs to blast fish to the surface. In Java and Bali, almost all the mangrove forests have been destroyed to make way for shrimp farms, rice fields, and tourist facilities. Along North America's Gulf Coast, tons of nutrient and chemical waste are swept into the ocean, poisoning shellfish beds. Coastal fisheries such as Canada's Grand Banks are closed as a result of overharvesting.

Coastal areas are staggering under an onslaught of human activity. Throughout much of the world, coastal zones are overcrowded, overdeveloped, and overexploited. Globally, we may already have lost half the Earth's coastal wetlands. And we are presently in the process of destroying 70 percent of the world's 600,000 square kilometers of coral reefs, an ecosystem containing some 200,000 different species and rivaling tropical rain forests in biodiversity. A combination of pollution, habitat destruction, and gross overfishing has led to the collapse of major fisheries and paved the way for malnutrition and disease in regions where people fish for subsistence.

Don Hinrichsen, *Issues in Science and Technology*, Summer 1996.

Even when there is some element of political will on the part of a number of the signatories to a convention, it is not always enough. Fifteen years after the International Whaling Commission voted in 1983 for an indefinite global moratorium on commercial whaling, for example, the IWC remains powerless to prevent Japan and Norway from killing hundreds of whales a year under the guise of "scientific research."

Even getting to the stage where strong international commitments to protect the ocean are put down on paper has been, thanks to inertia from governments and pressure from industry, far from simple.

In 1995, for example, representatives of nations from around the world gathered in Washington, D.C., and agreed to negotiate a treaty that would severely curtail production and emissions of persistent organic pollutants. Initially, says Boyce Thorne-Miller, the plan had been to work toward eliminating the tens of thousands of such pollutants in existence; it was finally agreed, however, to concentrate on only 12. And while these are all important contaminants—including PCBs, DDT and dioxins—many of them, Thorne-Miller says, "are no longer made in Western Europe or the United States, so it's not such a great hardship for the chemical industry to give them up. I overheard a member of one national delegation checking with an industry representative: 'This list OK with you guys?'". . .

"To be honest," sighs Mike Sutton, director of the Endangered Seas Campaign for World Wildlife Fund (WWF) International, "I've become so disappointed with the political process that I've begun moving away from the political scene altogether. I tend to doubt that the political process is going to get us where we need to be. The inevitable compromise between conservation and exploitation almost invariably tends to leave us in a position which does not provide the protection the environment needs."

None of which is to say that international conventions and agreements are without merit. Sutton agrees that "they need to get ratified and implemented." Boyce Thorne-Miller sees them very much as "a tool that we can use to bring pressure on governments and industry.". . .

But, not least because of the laborious nature of bringing an agreement to fruition and the considerable weaknesses and loopholes that even the best agreements almost invariably contain, environmentalists are increasingly looking at other means to bring about change.

CONSUMER MOVEMENTS

The WWF Endangered Seas Campaign, for example, has begun focusing more on the market—and, specifically, working with food giant Unilever to establish a Marine Stewardship Council, setting up a global, industry-wide mechanism for identifying and labeling sustainably caught fish. In India, the National Fishworkers' Forum is seeking to establish the first-ever international association of small-scale, inshore fishers, to draw global attention to the threat to their livelihood from giant offshore fishing fleets and the destructive environmental and social effects of shrimp aquaculture.

"I've never seen anything quite like the burgeoning opposition to shrimp aquaculture," says Greenpeace's international oceans campaign coordinator, Matthew Gianni. "It's a real grassroots movement, the thrust of which is trying to persuade American consumers—who, according to our research, eat more than 50 percent of the world's farmed shrimp—that 'all you can eat' offers from Red Lobster or whatever really aren't such good deals, at least not from the point of view of the environment or of inshore fishers in places like India, Bangladesh, Thailand and Ecuador."

Indeed, for many, that kind of effort—making consumers and citizens aware of the way in which their actions impact on ocean and coastal ecosystems, sometimes thousands of miles away—is the most important exercise of all.

DEVELOPING AN OCEAN ETHIC

As Sylvia Earle observes, maybe what we need is to develop an "ocean ethic"—a recognition that the ocean, far from being a "great matrix that man cannot sully and cannot appreciably despoil," an endless provider of resources or a bottomless sink for wastes, is as finite, and as vulnerable to human impacts, as any other environment. And the decisions that we all make—to build one more house near the coast, to drive a car when we could walk or take public transport, to eat one more plateful of shrimp—can all combine to the ocean's detriment.

"There are many unknowns," Earle admits, "but one thing is certain: we have the power to undermine the healthy functioning of the sea that supports us and all of the rest of life on Earth, but no sure way to heal the harm. For ages, the sea has taken care of us. For ourselves and all who follow, the time has clearly come for us to take care of the sea."

"In the long run, it is people's attitudes, prejudices, and perspectives that will influence the future of the oceans."

HOPE EXISTS FOR THE FUTURE OF THE WORLD'S OCEANS AND COASTAL AREAS

Michael Weber and Judith Gradwohl

In the following viewpoint, excerpted from the book *The Wealth of Oceans*, coauthors Michael Weber and Judith Gradwohl look into the future of the world's seas. They identify both positive and negative trends affecting the prospective development of the oceans. Weber and Gradwohl conclude that community activism and a growing public awareness of how human actions affect ocean health may help to preserve and even rebuild marine ecosystems. Weber is a freelance writer, conservationist, and former official of the National Marine Fisheries Service. Gradwohl is a museum curator with the Smithsonian Institution.

As you read, consider the following questions:

1. How have predictions about the ocean's future changed over time, according to the authors?
2. What do Weber and Gradwohl identify as positive trends that will affect the future of the world's oceans?
3. What examples do Weber and Gradwohl give of community activities designed to preserve the ocean habitat?

As we approach the end of the 20th century, concern about the abundance, diversity, and quality of life on earth mounts. Besides the gloom that feeds on predictions of global warming, of future losses in fisheries, wetlands, water quality, and jobs, people feel less and less a part of broader communities and find division rather than communion among themselves. Increasingly, people seem cynical about the institutions that they press to improve their lives. The gap between rich and poor has never been greater, nor affected so many.

So complex a situation seems to demand an equally complex response. Indeed, increasingly elaborate regulatory schemes, treaties, economic arrangements, and computerized models offer a convenient haven for business-as-usual. But intricate solutions generally are not as flexible and adaptable as the manifold uncertainty of the future demands. Furthermore, complex solutions generated by centralized government generally preempt discussion and agreement on the values and goals that motivate individuals and communities. Centralized solutions also often ignore important differences among regions, states, and communities that can help or hinder efforts to solve environmental problems. Narrow-interest politics and short-term gain regularly triumph.

In this viewpoint, we propose a way of thinking about the future and developing the ability to make the most of what will likely be many challenging situations. Recognizing underlying trends and barriers to change can increase the odds of making a transition to sustainable ways of life in communities of people that no longer destroy but instead work to restore ocean and coastal resources.

In the long run, it is people's attitudes, prejudices, and perspectives that will influence the future of the oceans, and of all of us who depend upon them. Already, economists, conservationists, business owners, community leaders, villagers, and citizens are fashioning tools for fostering the kind of changes that address the challenges confronting us and for seizing the opportunities arising from a reassessment of our place on the planet. But all the technical fixes in the world will accomplish little without a fundamental transformation in attitudes—from thoughtlessly taking all we can get toward thoughtfully living with what we need.

THINKING ABOUT THE FUTURE

In his 1966 book *The Challenge of the Seven Seas*, Senator Claiborne Pell of Rhode Island described his vision of the oceans in the year 1996. Besides envisioning people swimming underwater with artificial gills and communicating with dolphins, Senator Pell foresaw enormous aquaculture operations offshore, nuclear-powered submarine tankers, and underwater resorts. In Senator Pell's view of the oceans in 1996, an International Sea Patrol enforced fisheries management measures and monitored the ocean environment, including the waters around a new nation called Sidonia, which had been built atop a submerged mountain on the high seas.

Like other predictions of the future, Senator Pell's vision for 1996 reflects the attitudes of the times, which viewed the oceans as a new frontier for development. Several years later, the report of the congressionally chartered Stratton Commission shared Senator Pell's optimism about the potential for technology to tap vast resources in the sea. Among other things, the commission predicted that total annual fish landings from the oceans could mount to more than 550 million tons, six times current best estimates. These riches, it was thought, could help meet the growing global demand for food.

In recent years, these visions of future riches from the seas have faded. In developing a long-term strategy for the Rosenstiel School of Marine and Atmospheric Science at the University of Miami, Bruce R. Rosendahl recently described his own vision of the oceans in the year 2043. Instead of harvesting enormous quantities of fish from the wild, people produce food fish entirely by aquaculture. Yellowfin tuna, raised in large offshore pens, no longer are canned but are consumed as sashimi or sushi by the wealthy. Near cities, bays are so crowded that boaters are allocated boating days based on their registration numbers. Enormous hurricanes repeatedly strike the Atlantic and Gulf coasts, as well as Caribbean islands, devastating state economies. Together with payments for hurricane damage, payments for a devastating earthquake in the San Francisco Bay area destroy the insurance industry.

At United Nations meetings on high seas fisheries in July 1993, the environmental newsletter ECO published a fantasy of the oceans in the year 2052 that reflects a rueful view that life in the oceans will give way completely to industrialization. In a fictional interview, a chief executive officer boasts that his multinational corporation, called Living Marine Resources International, holds complete rights to the management of entire ecosystems and produces jellyfish protein packaged as fish. Tapping the tourist market, the multinational also breeds docile whales that people pay money to watch as a way of retaining "the illusion that the seas are some vast, wild frontier." Traditional fishing has become a matter for museum exhibits, and the corporation refurbishes commercial fishing boats to take tourists to former fishing grounds such as Georges Bank off New England.

In the last two decades, creating scenarios for the oceans of the future has become a more deliberate activity, aimed at providing decision makers with assessments of the consequences of their decisions. . . . Regional coastal programs have also developed computerized models to guide decision making. For instance, state and federal agencies have developed a computerized model for Long Island Sound that describes the actions and costs involved in reducing the flow of nutrients into the sound. Among other things, this model shows that because of projected population growth and changes in land use and transportation, currently planned reductions in nutrient discharges from sewage treatment plants will not be enough. The modeling has allowed state legislators, federal agency officials, town council members, and citizens to assess how much they are willing to spend to clean up the sound. . . .

OCEANS ARE RESILIENT

Indignation and concern are appropriate responses to the problem of coastal degradation. However, society must not be misled when the media leaves the impression that the oceans are hopelessly degraded by pollution.

The United Nations (U.N.) advisory group on marine pollution states without equivocation that "no area of the ocean and none of its resources appear to be irrevocably damaged.". . .

Marine systems are resilient. These ecosystems appear to have the capacity to rejuvenate, despite assaults on their integrity through high concentrations of pollutants and other human-induced insults.

R.L. Swanson, J.R. Schubel, and A.S. Weste-Valle, *Forum for Applied Research and Public Policy,* Spring 1994.

One need not be the head of a federal agency or a legislator to gain from even a little bit of structured thinking about daily life. Thinking how energy or water flows into and out of your house or relating your current and future use of gasoline to the release of carbon dioxide or nitrogen oxide that later fall on pavement or water bodies are just several examples of how to think holistically about the future at a personal level.

Similarly, we are all linked to other people in our own communities and to people in other communities. Examining your dependence upon communities upstream and downstream from your own community can reveal important relationships. Where is the sewage from your household discharged? Is your trash shipped to another community? How will converting your farmland to a residential subdivision affect the streams and coastal waters in your watershed? What food, energy, or services are produced in your community? What is currently imported from other communities, states, or countries?

Like other challenges to ocean conservation, thinking about the future and our relationships to those around us needs the attention of government agencies, businesses, and individual citizens. No one group has the answers or the tools to move us toward restoring the wealth of coastal and ocean resources. In the future, oceans and coasts will be more crowded, greater demands will be placed on marine waters and resources, and change will happen more quickly. We could afford to live on automatic pilot in the past. We must be more deliberate and thoughtful in the future.

KEY TRENDS FOR THE OCEANS

Setting the stage for future action can begin with describing trends or themes that are likely to shape future conditions and decisions. Some of these trends, such as population growth, already have gained momentum that we may slow but not reverse in the next several decades. Time itself and the mismatch between human institutions and natural processes form a second group of factors. A third group reflects common patterns of short-term thinking, which we can change. Likewise, we can acknowledge and try to compensate for our lack of knowledge about how the world works and our tendency to make only small changes when large changes are required.

Foremost among major trends is a continued growth in the number of people and their consumption of natural resources, from air and water to fish and forests. Even if birth rates fall much faster than they have in recent decades, the momentum from past population growth insures that the number of people will continue growing for decades to come. Supporting even the simplest of lifestyles will require converting more land from forests and wetlands into farmlands and cities, and will generate greater amounts of human, agricultural, and industrial wastes. Much of this waste will likely be discharged untreated into rivers and coastal waters. In coastal areas, population growth and related problems will intensify if the attractive life of cities continues to encourage people to move from inland communities to the coasts.

31

Like population growth, global warming and sea-level rise are trends already in motion. Even if the world's nations do reduce their carbon dioxide emissions to 1990 levels, as they committed to do at the Earth Summit in 1992, global warming has already begun. Most analysts believe that the only questions now are how much warmer the climate will get and what the secondary effects will be, such as sea-level rise in specific areas. These issues can seem abstract as long as you are not among the tens of millions of people living in low-lying coastal and delta areas.

The gap between consumption levels in developed and developing countries will tend to be a source of uncertainty and political anarchy around the world, reducing the ability for individuals, businesses, and governments to reverse wasteful use of resources and damage to the environment. During the last 40 years, the world economy expanded fivefold, while the number of people in absolute poverty and the gap between rich and poor doubled. Although they have fueled their own growth by relying on developing countries for expanding consumer markets and for raw materials such as shrimp and oil, most developed countries and multinational corporations have committed to only token efforts to reduce the gap. . . .

Other trends have to do with established patterns of thought and behavior that people and their institutions are unlikely to change, however desirable change might be. For instance, we expect that industrialized countries will continue to foster unsustainable use of coastal and ocean resources at home and abroad by promoting short-term thinking through foreign assistance programs, international trade rules, and bank lending.

Traditional foreign aid programs and lending by the World Bank and other multilateral banks reflect a model of economic and social development that measures welfare principally in short-term monetary values and assumes that a world of finite resources can support continued economic growth on a scale that will preserve the amenities that the industrialized countries enjoy while increasing the welfare of the developing world. Although recent administrations have embraced the theme of sustainable development that the 1992 Earth Summit sought to promote, they have held tenaciously to the goal of economic growth as the primary criterion of sound policy both domestically and abroad. As in the past, U.S. policy has assumed that we can grow our way toward a healthy future.

Among the consequences of this strategic error will be continued promotion and support for unsustainable use of natural resources and increasingly frequent collapses of fisheries and other coastal and ocean resources, including coral reefs, wet-lands, and water quality. Pressed to rescue species whose deple-tion reflects the collapse of economically valuable resources and the resulting loss of jobs, government will find itself fighting rearguard actions and being blamed for the casualties of eco-nomic growth. . . .

Finally, new technology and schemes for exploiting fisheries and other living and nonliving resources will be adopted with-out meaningful evaluation of their long-term ecological, social, and economic effects. As Paul Gray, the president of the Mas-sachusetts Institute of Technology, suggested in 1989, "New technology will be applied in ways that transcend the intentions and purposes of its creators, and new technology will reveal con-sequences that were not anticipated." Measuring success in terms of short-term benefits will obscure ways in which new technol-ogy may be used without undermining long-term benefits. . . .

POSITIVE TRENDS

Other trends offer hope that the trends just described will not dominate the future. Chief among these is the growth in public awareness and concern about the oceans that has marked the last several decades. Fed initially by dismay over the decline of great whales and dolphins, public concern about the oceans has grad-ually expanded to include the more visible forms of pollution, such as marine debris and oil spills, and most recently the ex-ploitation of fish and shellfish. The principal challenge of the fu-ture will be to further expand awareness so that people become concerned with less visible but more pervasive forms of pollu-tion, such as sewage, and the links between land activities and the oceans, including transportation and land use. This expan-sion will empower far more people to make a direct contribu-tion to the future, both by altering their own activities and by participating in local decision making about the use of wetlands or the disposal of waste.

The explosive growth in telecommunications, from comput-erized networks to Cable News Network, has made it possible to identify and respond to disturbing trends or destructive activi-ties sooner than in the past. Now, people around the world can learn about the violation of environmental standards in the con-struction of a major dam in India by reading on a computer network a report filed by an activist at the dam site. Thousands of letters may now pour into government offices in response to electronic bulletins that were unthinkable just five years ago.

Other technological developments promise to increase the

ability of governments to ensure compliance with conservation measures even on the high seas. By requiring the installation of satellite transponders in the early 1990s, enforcement agencies in the United States were able to track drift-net vessels hundreds of miles at sea and to determine whether the vessels were complying with the United Nations resolution on phasing out the use of drift nets. Now, two consulting groups in Seattle are trying to develop the means for tracking outlaw vessels, which do not carry transponders, by using satellite-based instruments to monitor the plumes of exhaust from their engines.

As increasingly sophisticated satellite instruments become available, monitoring areas or features as large as entire continents, oceans, and layers in the atmosphere or as small as acres of coastal wetlands or submerged reefs will become more common and, in some cases, more affordable. Rather than having to develop their own monitoring programs, the governments of developing countries and local governments elsewhere will be able to rely upon satellite information for surveying their jurisdictions. Similar developments in computer technology, including the graphic display of different kinds of information, will increase the ability of scientists to study entire systems and of decisionmakers to assess the effects of their actions.

Government officials and others are increasingly aware of the benefits of managing activities on the basis of ecological principles and are beginning to realize that the complexity and unpredictability of human society and ecosystems demands conservative approaches to exploitation of habitats and living marine resources. Similarly, the new field of ecological economics has provoked a needed discussion of the limits to current accounting for ocean and coastal resources, and has reintroduced some of the broader philosophical considerations that animated early economic thought. Already, the World Bank has taken some steps to evaluate its projects from a broader perspective that includes impacts on ecological, social, and cultural values. . . .

PROMOTING CHANGE

Just as the problems, threats, and trends that make up the ocean challenge arise from decisions by individual people, businesses, and governments, so too does the challenge to change apply at many different levels. For too long, we have failed to ask ourselves the question: Who can best promote the changes that the future demands? The long-standing formula calls for policy debates, legislative hearings, legislative compromises, unfunded national and international standards, fundraising campaigns, and more meetings. This pattern results not simply because policy makers inside and outside of government pursue the well-worn path of seeking uniform solutions at the highest possible level. The pattern arises as well because individuals and communities prefer to remove local or regional controversy and problems to higher levels of government, thereby avoiding the difficult decisions that will actually affect people's lives.

There is no clearer example of this pattern than the collapse of many coastal fisheries and the decline of fishing fleets. Although the regional fishery management councils established by the 1976 Magnuson Fishery Conservation and Management Act arguably were meant to promote greater compliance with fishery regulations through involvement of regulated fishermen, they often have served as a means for fishermen to avoid managing themselves and, at the same time, for communities to continue marginalizing commercial fishermen by arguing that the health of the commercial fishing industry is a federal matter. It is as if the Magnuson Act prohibited fishermen themselves from taking steps to maintain and restore public fishery resources, which they benefit from using. Now, decisions affecting fishermen are taken up at levels farther and farther removed from the water, and a vicious cycle of avoiding responsibility and castigating government has set in.

One opportunity in this common situation is to redirect the time, money, and effort that continually fuels policy debates in Washington toward fostering a desire and capability for self-determination and cooperation among fishermen and coastal communities. At a minimum, federal and state governments should not mask the depletion of resources through subsidies or other mechanisms; national standards and management programs should insure that the interests of the general public in the conservation of fisheries are met. . . .

We can greatly improve our ability to maintain and restore the productivity and diversity of marine and coastal areas also by insuring that, like the ecosystems that we depend upon, we change management measures in response to changed circumstance. Too often, laws are passed or decisions taken and forgotten until the problem they were meant to address or new, unanticipated problems reach crisis proportions. Most regulations, laws, programs, and actions that governments, businesses, and people take are based on a theory about how the world works. Paying attention to the way in which fish populations, water quality, or our communities respond to laws or actions can increase the effectiveness of what we do. Being able to track the effectiveness of our actions, however, will depend upon better and more thorough monitoring, not just of water quality and wildlife populations, but of the movement of people and changes in economic activity. . . .

If ocean conservation emerges from its isolation, the opportunity for gaining strength through collaboration with other people and institutions will grow. Certainly, those interested in the conservation of coastal forests have much to offer those concerned with protecting coastal waters from runoff and anadromous fisheries from loss of spawning habitats. Similarly, the interests of those concerned about freshwater quality have much in common with those concerned with marine water quality. Learning to relate ocean concerns to the concerns of other interests will require broadening perspectives and valuing generalists more than has been the case in the past.

Similar opportunities exist among ocean professionals. Pooling the perspectives and interests of economists, community development specialists, anthropologists, fishery biologists, water quality specialists, and coastal planners, just to name a few, offers the potential for broadening the constituency for specific issues that now suffer from narrow support. . . .

COASTAL COMMUNITIES

Just as watersheds define the physical relationship between land and sea, so too do communities define the relationship between people and the oceans. Learning to respect and support this relationship may be one of the most important contributions that national governments can make to the future of conservation. Already, communities are creating opportunities for this kind of partnership by recovering for themselves the responsibility and power that they had given to others.

Over a period of two years, more than 200 citizens in Puget Sound proposed, discussed, revised, and adopted 40 measures of the health of their community. In this intensely democratic process, citizens struggled to describe what future they would like to see in their region. Sustainable Seattle tried to capture the future as a whole rather than as a conglomeration of particular interests. For these citizens, a booming economy would not be enough, if the environment were being wasted, crime were rising, and infant birth weights continued to fall. They were convinced that one key to restoring their community was actively recognizing and probing the links among seemingly unrelated aspects of community life.

No measure of the community's future captured the imagination of these citizens more than the abundance of wild salmon in local streams. Revered by Native Americans as one link to the earth, Pacific salmon has been a source of pride for the people of the Northwest for many years. Unlike hatchery-raised salmon, wild salmon depend upon healthy streams for spawning and early growth. Degradation of water quality or the streambed, such as often is caused by development of surrounding lands, reduces successful salmon reproduction and eventually leads to a decline in the size of salmon runs.

Sustainable Seattle found that the salmon runs in the two local creeks they had chosen as indicators had declined steadily for at least the last 15 years. In their 1993 report to the community, Sustainable Seattle put the losses in a broader context. "Our careless stewardship of salmon—perhaps the most symbolically and economically important creature to share Puget Sound with us—is reflective of our attitude toward a variety of living systems, from neighborhoods to ecosystems."

Drawing support from national and international organizations as well as other local groups, Sustainable Seattle is just one of dozens of citizen groups that are not waiting for change to occur to them, but are launching themselves into an experiment with the future. In the end, they are attacking two of the gravest threats to oceans and coasts, to forests and wetlands, to children and healthy communities: indifference and isolation.

On the other side of the Pacific Ocean, in Bais Bay on the east coast of the Philippine island of Negros, residents in the village of Bindoy already have moved from studying to restoring their lands and waters. In the mid-1980s, collapsing fisheries and the obliteration of local mangrove forests and coral reefs convinced Wilson Vailoces to seek the help of Silliman University in reversing the decline. By 1993, after forming a voluntary group called Sea Watchers, Wilson and others from the village of Bindoy had planted more than 100,000 mangrove trees and had sunk more than 1,000 artificial reefs made of bamboo, tires, or concrete. The Sea Watchers did not stop with recreating habitat for fish, but convinced farmers to reduce their use of pesticides and to avoid tillage practices that send soil into coastal waters. Fifty members of Sea Watchers now patrol local waters to enforce restrictions on the use of dynamite, poison, and fine-mesh nets, leading to fines for hundreds and prison for a few.

No longer are the people of Bindoy simply exploiting their surroundings. They are rebuilding them. Their journey toward restoration of the environment that supports them began with the dream of Wilson Vailoces to build a future for his community.

It's a dream that can work again.

|"The oceans, which once seemed
limitless, can no longer sustain the
demands placed on them by market
forces to provide fish [and] other
marine life."

THE WORLD'S OCEAN FISHERIES ARE
IN A STATE OF COLLAPSE

Barry Kent MacKay and Fran Stricker

In the following viewpoint, Barry Kent MacKay and Fran Stricker argue that fish populations around the world are in a state of serious decline due to overfishing and ocean pollution. Commercial fishing is destroying marine habitat and is depleting fish stocks faster than they can reproduce, the authors contend. They note that some fisheries, such as the cod populations in the Atlantic Ocean off of Canada, have collapsed entirely. MacKay and Stricker are staff writers for *Mainstream*, a publication of the Animal Protection Institute, an animal rights organization.

As you read, consider the following questions:

1. Why do the authors consider deep-sea trawlers to be an especially serious threat to ocean life?
2. How much marine wildlife is caught and then discarded as "bycatch" each year, according to the United Nations Food and Agricultural Association?
3. What are the effects of agricultural runoff and global warming on fish populations, according to MacKay and Stricker?

Excerpted from Barry Kent MacKay and Fran Stricker, "Something Fishy Going On," *Mainstream*, Spring 1998. Reprinted by permission of the Animal Protection Institute.

There are 30 million species of animal in the world, and 99% of them live in the world's oceans. Yet close to 99% of animal protectionists' attention, compassion, and protective endeavor is paid to the 1% that live on land. The few exceptions are marine mammals—whales, dolphins, porpoises, seals, sea lions, and others—warm-blooded, demonstrably intelligent, forming social bonds. Despite the horrid abuses we heap on them, they are enough like us fellow mammals to attract efforts to cherish and protect them.

However, of the remainder of marine life—even if we delete the huge majority who have rather poorly developed nervous systems and subsequent intellectual capability, or who simply lack central nervous systems altogether so perhaps we need not worry that they suffer—we are certainly left with one group that receives very little attention from the animal protection movement.

Fish.

Fish feel pain. Whatever their intellectual limitations, they are most certainly capable of suffering. And they do so by the multi-billions in the interest of producing profits for humans.

Most fish species produce huge numbers of young to offset correspondingly high natural mortality rates. Species with such high "recruitment rates"—of millions of eggs, only a handful will produce fish that survive to sexual maturity—are noted for their ability to "bounce back" from extremely low numbers. Put simply, the more you remove from the population, the greater the survival potential for those that survive, thus "overfishing" is unlikely to be a problem. Indeed, some species, such as the herring, have been considered by fishery biologists to be immune to endangerment through overfishing.

But they are not. At least 60% of the world's 200 most commercially valuable fish species are either overfished or fished to the limit. Fish stock after fish stock has plummeted into precipitous decline. Our insatiable appetite for neatly packaged fish sticks at the supermarket, for salmon sandwiches, fish and chips, swordfish steaks, tunafish catfood, pizza with anchovies, and genuine sushi has taken a fearful and continuing toll. Tempers grow thin as, far too late, governments seek to implement protective measures. As countries compete for dwindling international stocks on the open seas, each blames the other for the declines.

THE NORTHERN COD STORY

Origins of the problems are complex, and perhaps best illustrated by the plight of the northern cod stocks of the Northwest Atlantic fishery. "King Cod," the fish that "built" Boston, was unimaginably abundant when Europeans first reached our Atlantic coast, about 1,000 years ago. In 1497, John Cabot described the seas over the Grand Banks, off Newfoundland, as so "swarming with fish [that they] could be taken not only with a net but in baskets let down [and weighted] with stone."

Closing the cod fishery throughout most of Newfoundland and Labrador in 1992 delivered a severe economic blow to a remote and harsh area of chronic low employment opportunity. After such abundance, how could this come about?

The Canadian government has blamed seals, Spanish fishermen, and climate changes, but the simple answer is greed.

In *New Scientist*, September 16, 1995, Deborah MacKenzie summarized "the disaster of the Grand Banks" as "a compendium of the mistakes being made by fisheries all over the world." In blunt detail she related that "when scientists began to manage the Banks in the 1950s, they promised to assign 'safe' quotas to Canadian and foreign fleets. They failed."

FISHING'S TROUBLED FUTURE

What is the future for fish and fishing in the world's oceans? Clearly it's grim if current levels and methods of exploitation continue. With their own coastal fisheries depleted, the United States, European countries, and Japan now compete to import food fish from developing nations, where fisheries are even more poorly managed. According to the United Nations' Food and Agricultural Organization (FAO), 11 of the 15 main marine fishing grounds are seriously depleted. As a result, fishermen must expend more effort to hunt down fewer fish.

Wesley Marx, *California Coast & Ocean*, Winter 1997–98.

They failed because the scientists' data were inaccurate—taken from random samplings, whereas the commercial fleets went directly to where they knew the cod would congregate—and they would not abandon the basis of their theories on how quickly the cod would recover their populations. The cod catch continued to fall, from 810,000 tonnes in 1968 (in the United States, a "ton" is 2,000 pounds; in most of the rest of the world, a "tonne" is 1 million grams, or 1,000 kilograms, and is equivalent to 1.1 U.S. tons) to 150,000 tonnes by 1977. The "total allowable catch" (TAC), which scientific theory said should allow fish stocks to increase, and which was set by the Canadian Department of Fisheries and Oceans (DFO) at about 16% of the fish population, never seemed to work. Despite enormous pressure from the big fishing companies to raise the TAC, DFO kept lowering the TAC, then in June 1992 "recommended banning fishing altogether," wrote MacKenzie. "Suddenly, the scientists realized there were no cod old enough to spawn left.

"By now the fishermen were worried too, and agreed to a fishing moratorium on the Bank and adjacent fisheries. In 1993, it was extended indefinitely."

Since then media attention has belatedly focused on the role politics played in the disastrous decisions made by DFO. The politicians who ran DFO, intent on pleasing the powerful multinational corporate players in the fishing industry, ignored their scientists' decade-old warnings of overfishing.

Not an Isolated Incident

This destruction of one of the world's richest fisheries is not an isolated incident. It's the same the world over. Greed and increased use of sophisticated technology are the root of the problem. The oceans, which once seemed limitless, can no longer sustain the demands placed on them by market forces to provide fish, other marine life, and other resources.

Even when conservation is recognized, fisheries are impossible to manage. Fishermen have dumped numbers over quotas, or catches of protected species, rather than face fines for illegal catches. Catch sizes have been under-reported, or not reported at all in a high seas free-for-all.

Deep sea trawlers are particularly destructive. Huge nets are dragged across the sea floor, destroying breeding habitats and feeding grounds while sweeping up every living thing in their paths. This "mining of the oceans" damages ocean communities and disrupts complex food chains.

Factory ships process the catches at sea, with no need to interrupt the continuing slaughter. Sophisticated devices remove the guesswork. The fish have little opportunity to escape detection and subsequent destruction. Long floating nets, the notorious "drift nets"—sometimes stretching 30 miles or more—also catch and kill all species, indiscriminately.

A History of Waste

The history of fishing is a history of waste. For example, in the North Atlantic a small fish called the capelin, one of several species that is a significant part of the food chain, eaten by whales, seabirds, and various larger fish species, occurs by the billions. Its numbers seem indestructible, such that fishermen used to simply dump them, or possibly use them as fertilizer. But with declines in other, more profitable fish stocks, markets have opened for this small but vitally important fish. Japan, a fish-eating nation with endlessly accelerating demands for seafood, expressed willingness to buy the roe (eggs) of the capelin. Fishermen complied, harvesting only the eggs, and discarding the rest of the females' bodies, plus, of course, all of the males.

Some researchers attribute recent epidemics of starvation of Atlantic puffins and other North Atlantic seabirds on a decline in capelin. The great schools of capelin, once thought to be inexhaustible, are no longer a dependable occurrence.

In far warmer waters, in the Gulf of Mexico, the red snapper population has been reduced not only by overfishing, but by the "incidental" catch of shrimp fishermen who kill and waste about 12 million young red snappers annually. It's been estimated that, for each pound of shrimp, shrimp fishermen catch about ten pounds of fish, most killed and discarded. Enjoy your shrimp cocktail—it came at a formidable price.

The United Nations Food and Agriculture Organization (FAO) estimates that 12 to 20 billion pounds of marine wildlife is caught and dumped each year as a "bycatch." In that vast volume of waste are numerous dolphins, sea turtles, albatrosses, shearwaters, sea lions, and other non-fish species. The wasted bycatch is equivalent to 10 pounds of food for every person on Earth. The bycatch is not limited to fish. In the Bering Sea, the area with the world's highest bykill (9 million tonnes), in 1992 a discarded 16 million red king crabs exceeded the target catch of about 3 million. . . .

In 1994 the FAO's surveys showed that of the world's 17 leading fishing grounds, 13 were depleted or in severe decline. Shortly afterward Norway and Iceland fell into bitter dispute over access to fisheries around the Arctic archipelago of Svlaboard. Norway fired warning shots and cut the nets of Icelandic trawlers, who eventually were forced to leave. And it was reported from the vast inland Black Sea of Asia Minor, that of the 26 fish species historically caught in the Black Sea, only 5 remained. In less than a decade fish catch had gone from 700,000 tons per year to 100,000. Here it was not just overfishing, but massive pollution generated by the 160 million people living in the Black Sea drainage that had contributed to the destruction. . . .

SALMON IN DECLINE

Among the most famous North American fish declines are the losses in west coast salmonids. Overfishing is again a culprit, but not the only one. The salt water salmon breed in fresh water streams and the newly hatched baby salmon make their way downstream to the sea, where they spend several years growing to sexual maturity. In a final act the mature fish then make the perilous trip back upstream to the place of their birth, to reproduce and die.

But dams block the spawning salmon, and logging practices fill those streams with pollution and debris hazardous to the fish or leave fragile eggs exposed to harmful levels of sunlight. Agricultural runoff invariably makes its way to inland river systems. And, of course, there are the fishermen. Salmon numbers are in serious decline. . . .

POLLUTION ENDANGERS FISH

Because they are unseen, hidden from our awareness by the reflective surface of their aquatic home, we spend little time thinking about our effects on fish. Approximately 3.6 billion people, more than half the world's human population, live within 100 miles of the coast. And nearly all of the 5 billion-plus of us now living depend on technologies that generate voluminous waste, much of it toxic and most of which ultimately makes its way down to the waterways and the seas. The rivers, lakes, and oceans are vast dumping grounds fed by rivers that have become sewers for our civilization. Dr JoAnn Burkholder of North Carolina State University said, "It's hard to imagine that farming on land and building in cities could harm the marine environment and fishermen, but it does. The tons of sewage produced by millions of people don't just go away when we flush . . . a lot of it winds up in our coastal waters. And construction, agriculture and logging send clouds of choking sediments and excess nutrients into marine waters, smothering sensitive habitats. What we do on land profoundly affects life in the sea."

Food chains fundamental to fish survival often originate in complex shoreline marshes, mangrove swamps, and other habitats continually degraded and destroyed in the interest of short-term economic gain. Coral reefs, also fundamental as homes to numerous marine organisms and nurseries harboring the beginnings of various elaborate food chains that support so many more forms of wildlife, and humans, are being destroyed by siltation, pollution, and destruction for harbors and marinas. Indeed, corals are being lost directly to the aquarium trade in response to market demand for "living rock." The fresh water that flows from the land becomes ever less as demands for irrigation siphon it off, altering salinity levels in the nearshore waters. The mighty Colorado River, carver of the Grand Canyon, is just about entirely used up by the time it reaches the Sea of Cortez.

Fish die by the millions as a direct result of agricultural runoff, one form of food production destroying another. Non-native species are often intentionally or accidentally introduced, to compete with and often destroy the specialized native species and fracture delicately balanced interrelationships forming unique ecosystems. . . .

GLOBAL WARMING

While tropical waters produce the greatest biodiversity, colder water contains more oxygen and thus, on average, far more fish. Global warming, with its ability to change currents and water patterns and to melt polar ice, which will raise water levels and decrease salinity, poses an overwhelming threat to fish stocks already hammered in so many other ways. We don't know what effect reduction in the ozone layer may have, but we do know that many ocean food chains begin with the phytoplankton that spends part of the day near the sea's surface, and that there is a risk that increased ultraviolet solar radiation could seriously affect these tiny organisms. Already there are reports from the higher latitudes of the southern hemisphere of whales suffering sunburn, presumably because of increased solar radiation. We can warn humans to wear sunblock and sunglasses, but we can't do that for the rest of the world's species.

These huge risks may seem beyond our ability to even understand, let alone control. But we can safely say that we are stressing both freshwater and marine environments beyond anything that makes sense either from the standpoint of the animals thus victimized, or our own self interests. We are ultimately no less dependent upon a healthy, viable environment than any other species. But still we continue, and something has to give. The fish, like so many other species, are what is giving . . . what we are losing, as we so callously, and thoughtlessly, destroy them.

| *"All evidence . . . indicates that the ocean is relatively healthy."*

THE WORLD'S OCEAN FISHERIES ARE NOT IN A STATE OF COLLAPSE

Michael Parfit

Michael Parfit is a writer for *National Geographic* and *Smithsonian* magazines. In the following viewpoint, he argues that while problems exist in some ocean fisheries, the oceans in general are in relatively good shape. He contends that environmental organizations often exaggerate the extent of crises such as the collapse of ocean fisheries in order to attract the public's attention and gain support for their proposals. This emphasis on catastrophes does little to educate people about the state of the environment or advance long-term conservation goals, he concludes.

As you read, consider the following questions:

1. What nineteenth-century theory is evident in current environmental arguments, according to Parfit?
2. What does Parfit identify as the main long-term problem concerning the world's fisheries?
3. What benefits have environmental groups received from focusing on catastrophes, according to the author?

Reprinted from Michael Parfit, "Disasters Aren't the Problem," *The Washington Post*, December 17, 1995, by permission of the author.

B efore the dawn of modern geology in the 19th century many people believed the earth had been formed by a series of catastrophes. The idea was that incredible events—huge volcanic blasts, floods, violent winds—swiftly built our planet. That theory, now discredited, is known as catastrophism. Today a similar theory seems at work in environmental politics: Catastrophes are making the world fall apart. This approach is as wrong as the first one, and is encouraging the planet's decline even as it seeks to prevent it.

Recently I spent well over a year studying one of those alleged catastrophes—the state of the world's fisheries—for a *National Geographic* magazine article that appeared in the November 1995 issue. I began armed with environmentalist reports that claimed fisheries were collapsing worldwide. But I couldn't find such a disaster.

I pursued the story to New England, Alaska, Canada, the Philippines, Patagonia, the Falklands, South Africa, Spain, Iceland and Senegal. I found real trouble in some places, like the Grand Banks [off Newfoundland] and West Coast salmon fisheries. But in other places stocks were recovering because of tightened regulation and enforcement.

A COMPLEX SITUATION

What I saw was less immediate and more complex than a catastrophe. It is true that fishermen and their equipment have become dangerously powerful. But all evidence, including that from the United Nations' Food and Agriculture Organization, indicates that the ocean is relatively healthy. The processes that create food in the sea are largely intact. The main problem is long-term: We have to learn how to give up ancient freedoms to fish the sea without restraint—as many are already learning how to give up hunting at will, overgrazing and logging beyond forest growth.

Most of our ruination of the planet proceeds with similarly small events—the spraying of lawns; the dumping of oil; the demand for wormless apples, beachfront cottages, shrimp and layers of packaging—in which we all partake. Indeed, most environmental problems are far more accurately described by the newer theory of "uniformitarianism," which holds that the earth was made by the same forces we see today—the trickle of rain, the polishing of wind, the creep of continents—working steadily through unimaginable depths of time.

The most frightening thing I encountered in my research was not the cod crash off Newfoundland, but something a skipper told me in Dakar, Senegal: "The fish just get a little smaller each year." Like rain taking a mountain apart grain by grain, the unspectacular change takes the biggest toll. Yet much of the present environmental movement is as breathless as the early theory; Vicious Minerals Inc. is always on the verge of stripping the Last Great Wilderness. Partly this is because looming disaster is galvanizing. It's dramatic, motivating and hopeful to imagine that threats to the earth are crises caused by villains we can vanquish.

MEDIA HYPE

Clearly, there are some problems associated with the management of our nation's fisheries, but media hype over the past few years has overdramatized the problem by asserting that "our oceans are being emptied." Sensationalist stories, combined with the alarmist campaigns of environmental organizations, suggest that all ocean fisheries are being decimated.

Such portrayals of depleted fish stocks are unfair and inaccurate. Fisheries that produce the bulk of our seafood are producing fish year after year at sustainable levels. Take Alaska, for example, where record runs of salmon and billions of pounds of groundfish continue to be harvested each year.

Richard Gutting, *Forum for Applied Research and Public Policy*, Summer 1996.

But does it matter if people are misled? Exaggerated advocacy, as even Calvin and Hobbes know, seems to be the only way to keep people awake. Like the Pentagon's budget-time assessment of enemy weaponry, maybe this approach is necessary to engage our awareness before it is, as they say, too late. It was, after all, only the dramatic discovery of the spreading ozone hole that led to agreements to reduce chlorofluorocarbons.

THE PROBLEM OF CATASTROPHISM

Possibly. But it is also possible that environmental catastrophism defeats its own purpose.

With all the other pleas for righteousness out there, catastrophism is becoming easy to ignore. And it can be so inaccurately alarmist that it gives Vicious Minerals the upper hand in appealing to people of balance. Ozone depletion was predicted long before the hole was found, but because the public had become accustomed to excess, industry was able to tar the scientists who warned of it with a Greenpeace brush.

Most important, apocalyptic battles like this also obscure the point that human survival depends on universal and individual respect for all parts of the earth, pretty or not. Though environmental problems are long-term and caused by all of us, catastrophism encourages only short-term, narrow action and, by giving the impression that only a few bad guys are to blame, allows the rest of us to go on with our piecemeal destruction.

Take the fisheries case. Alarms about the need to pass a United Nations fishing resolution [regulating fishing on the high seas] called the measure "the last chance to avert collapse of world fisheries." The measure, aimed at the relatively few fishers who work in international waters, has been passed; the tide of scary stories ebbs.

Yet this last chance did not save fisheries. It was, in fact, a tiny step in what needs to be a widespread change in attitude. Like the rain forests, the ozone hole, and the waste of energy, other stories which have passed their catastrophic prime but remain important, the issue of fisheries continues. But environmentalists have shot the wad: The emergency stories have been done, the anxiety expended, the hour passed.

Environmental groups have thrived on catastrophism. It fills coffers and defines successful battles. Yet the war is being lost: slowly, relentlessly, the world grows more damaged. To fight this, environmentalists have urged corporations to become more cautious and farsighted; maybe they should take their own advice.

"Low-lying islands and intensely
developed coastal areas ... face an
urgent need to develop strategies for
coping with sea-level rise."

SEA-LEVEL RISE CAUSED BY GLOBAL WARMING IS A SERIOUS PROBLEM

Hans Hanson and Gunnar Lindh

The effect of carbon dioxide and other greenhouse gases on the world's climate has become a controversial environmental topic. Many people fear that the increased quantities of these gases produced by industrialization are creating a global warming trend. In the following viewpoint, Hans Hanson and Gunnar Lindh contend that rising temperatures may increase the volume of ocean water, melt glaciers, and cause sea levels around the world to rise. Ocean level increases of just a few inches in turn would create numerous social and environmental problems, they argue, including coastal erosion, flooding of populated areas, and saltwater intrusion into freshwater systems. Hanson and Lindh are professors of water resources engineering at the University of Lund in Sweden.

As you read, consider the following questions:

1. How many of the world's people live in coastal regions, according to Hanson and Lindh?
2. What is the Intergovernmental Panel on Climate Change (IPCC), according to the authors, and what has it predicted about sea-level rise?
3. What four possible long-term responses to rising sea levels does the IPCC outline, as reported by the authors?

Reprinted from Hans Hanson and Gunnar Lindh, "The Rising Risks of Rising Tides," *Forum for Applied Research and Public Policy*, Summer 1996, by permission of the authors. Endnotes in the original have been omitted in this reprint.

More than half the world's population—or about 2.5 billion people—live along ocean coastlines, making these areas the Earth's most densely populated and rapidly expanding regions.

In fact, about 40 percent of the U.S. population lives within 50 miles of the coast. Worldwide, 14 of the fastest growing urban centers are coastal cities.

Though these coastal zones represent the world's most economically and socially attractive locations, they face some of the most vexing environmental problems. Chief among them are severe coastal erosion, flooding, and saltwater intrusion into soil and freshwater rivers, all of which may be linked in part to sea-level rise. Though the actual cause of sea-level rise remains a subject of debate, many scientists blame global warming.

GLOBAL WARMING AND SEA-LEVEL RISE

There are several potential links between global warming and sea-level rise. First, temperature increases can cause the upper levels of the ocean to expand, increasing the water's volume. Second, increasing global temperatures can melt nonpolar mountain glaciers as well as the Greenland and Antarctic ice sheets. As these frozen regions thaw, their melt-off spills into the oceans.

Though no one can predict with certainty the amount of sea-level rise that will occur, the Intergovernmental Panel on Climate Change (IPCC), an organization established jointly by the World Meteorological Organization (WMO) and United Nations Environment Programme (UNEP), estimates a possible sea-level rise between now and 2030 of 3 to 11 inches (8 to 29 centimeters).

By 2070, the panel warns, the rise might be in the range of 8 to 28 inches (21 to 71 centimeters). Such forecasts are significant when compared to the 4- to 6-inch (10- to 15-centimeter) rise in sea level that has occurred over the past 100 years.

COASTAL DEVELOPMENT

Development-related activities along the world's coastlines—including housing construction, tourism, and recreation—only exacerbate the problems related to sea-level rise, because the closer human activities and investments move toward the coastline, the more sensitive it becomes to erosion and inundation.

Furthermore, because of the value placed on coastal development, tensions often surface in coastal communities between economic goals and conservation efforts aimed at protecting the coastal region's ecology.

Few public officials have developed comprehensive strategies to address the potential consequences of sea-level rise. To date, piecemeal approaches to the problem often have helped one area at the expense of another.

In Britain, for example, flood barriers constructed along the Thames River have protected the London metropolitan area from flooding. These barriers, however, have channeled the bulk of floodwaters downstream, thus placing areas east of London at greater risk during times of high water. In other cases, communities have shifted sands from healthy beaches to those affected by tidal erosion. In the process, they have put the donor beaches at risk.

MATTER OF DEGREES

How much will global mean temperatures rise, and how will the increase affect sea level? These are the critical questions that scientists face when trying to assess the impact of global warming on coastal regions.

Various scientific studies have led to different predictions of future temperature increases, but a rise of about 5 degrees Fahrenheit (3 degrees Celsius) is expected to occur over the next century, compared to a rise of less than 1 degree Fahrenheit (0.6 degrees Celsius) over the past hundred years.

Though many scientists concur that a 5-degree increase might occur over the next century, they still struggle to interpret what such an increase might mean, particularly if the increase occurs regionally rather than globally.

Consider that temperature increase varies widely among various parts of the globe. In the winter months, for instance, the temperature in the higher latitudes of the Northern Hemisphere may be 50 to 100 percent greater than the global mean. Such natural variability from region to region complicates analyses and fuels the arguments of those who contend that global warming is a figment of scientific imagination.

In fact, skeptics of global warming tend to see only natural environmental variability where other scientists see a demonstrable global temperature rise. This disagreement over the extent— or even the existence—of global warming makes planning— even short-term planning—difficult. Indeed, while scientists think in terms of 50 to 100 years, policy makers think in a much briefer frame, usually with an eye on the next election.

COASTAL CONCERNS

Though even a modest rise in sea level will affect all of the world's coastlines in countless ways, the most important consequences include the following:

• *Beach Erosion.* About 70 percent of the world's sandy coasts have suffered erosion in the past few decades. If sea-level rise continues, such losses will increase. Seaside resorts, which include some of the world's most valuable and intensely developed land, are especially vulnerable. Here, the high costs of land have resulted in the construction of high-rise buildings in close proximity to the ocean. Thus, the slightest shoreline retreat would endanger the continued safety of these structures.

• *Inland Flooding.* Inundation of inland fields also may accompany a rise in sea level. For example, according to one 1998 study, coastal strips as wide as 12.5 miles may become inundated if sea levels were to rise three feet. In the United States, the low-lying Atlantic and Gulf coasts are relatively flat and therefore most vulnerable to inundation.

Meanwhile, in the tropical storm zone of Southeast Asia, a 5-foot rise in the sea level of Bengal Delta could spell a loss of nearly one-fifth of the zone's habitable land.

• *Saltwater Intrusion.* As sea level rises, saltwater likely will intrude into coastal freshwater systems and leach into the soil. Indeed, higher sea levels will push saltwater into the ground through flooding or drive it upstream in coastal rivers.

Saltwater is particularly dangerous to agriculture, where fields and crops lie just above the water table. When sea level rises, the water table does the same, salinizing the soil and damaging or killing salt-intolerant crops. Removing salt from soil is expensive and difficult and would be a waste of money and time if the lands were to become resaturated with saltwater after they had been treated.

Saltwater may also intrude into wells, contaminating drinking water and forcing residents to purchase expensive desalinization equipment to render their water potable.

Furthermore, urban storm drains and sewers will likely be damaged by saltwater and will require expensive reconstruction and protective maintenance.

TURNING BACK THE TIDE

Clearly, the combined effects of erosion, inundation, and saltwater intrusion call for planning and implementation of remedial action.

Yet, the threat of sea-level rise requires a plan aimed at broad, long-term coastal-zone management. To that end, the Intergovernmental Panel on Climate Change recommends that coastal nations implement comprehensive coastal-zone management plans by 2000.

Many countries, however, especially developing nations, will find that charge difficult—perhaps impossible—to execute. The first and most daunting barrier is cost. The expense of turning back the tide likely will prove prohibitive for small-island developing countries, such as the Seychelles Islands or the Maldives, whose governmental budgets are minuscule compared to that of the United States.

For a 3-foot rise in sea level over the next century, mitigation costs are likely to exceed 1 to 3 percent of current gross domestic product for these island nations.

Beyond cost is the issue of institutional structure. Small developing countries are unlikely to have the sophisticated governmental agencies and programs needed to plan and implement short- and long-term coastal management plans.

Though these nations may lack the money and institutional structure to contend with the rising sea, lack of action will leave them awash in a host of problems that will probably grow more dramatic with time.

However, before these island nations can plan and implement mitigation efforts, they must accurately assess their vulnerability to sea-level rise.

COASTAL MANAGEMENT

For help, they might look to the Coastal Zone Management Subgroup of the Intergovernmental Panel on Climate Change Response Strategies. This subgroup should facilitate the efforts of coastal and island states to develop and, where possible, to implement coastal-zone management plans by 2000.

In 1991, this organization developed recommendations for (1) identifying and assessing physical, ecological, and socioeconomic risks posed by accelerated sea-level rise and other coastal impacts associated with global climate change; (2) assessing the impacts of development and other socioeconomic factors in influencing vulnerability to sea-level rise; (3) evaluating the effects of mitigation measures; and (4) measuring the capacity for integrating a worldwide response to sea-level rise into a broader coastal-zone management strategy.

The organization also outlined four possible response strategies: no response; retreat (gradual evacuation of human populations); protection (construction of seawalls and dikes); and accommodation (imposition of regulations that strictly control development and land use).

Though the long-term effects of sea-level rise remain to be determined, there is little doubt that rising oceans will affect much more than the world's coastlines. Indeed, the economic and demographic effects likely will be felt around the globe.

Nevertheless, low-lying islands and intensely developed coastal areas—places that will be affected first—face an urgent need to develop strategies for coping with sea-level rise. The time to begin is now.

"The much-ballyhooed climate models that predict global warming cannot make any quantitative predictions at all about sea levels."

THE THREAT OF RISING SEA LEVELS CAUSED BY GLOBAL WARMING IS UNPROVEN

S. Fred Singer

S. Fred Singer is a scientist and author whose books include *Hot Talk, Cold Science: Global Warming's Unfinished Debate*. He is founder and president of the Science and Environmental Policy Project, a nonprofit policy research group. In the following viewpoint, Singer takes issue with those who argue that global warming will cause ocean levels to rise and create environmental and social problems. He argues that the link between global warming and rising sea levels has not been proven. Global warming, if it exists at all, may actually cause sea levels to decline, Singer maintains, by stimulating greater precipitation in polar regions and creating higher concentrations of ice and snow in those areas. He calls for more scientific research of the problem.

As you read, consider the following questions:

1. Why are changes in sea level hard to measure, according to Singer?
2. According to Singer, what hypothesis is supported by climate and ocean data from the 1900–1940 period?
3. What steps should the United States and other industrialized nations *not* take in response to predictions of rising sea levels and global warming, in the author's judgment?

G lobal warming devotees have been making alarmist predictions about the rising sea levels they think will follow an increase in the earth's average temperatures. The horror stories include the flooding of low-lying coastal areas, the disappearance of island nations, the inundation of America by environmental refugees, and an exponential explosion in insurance claims. Activists apparently don't realize that the much-ballyhooed climate models that predict global warming cannot make any quantitative predictions at all about sea levels.

They also don't realize that informed speculation about rising sea levels has been steadily falling. Initial estimates by the Environmental Protection Agency projected that a doubling of atmospheric carbon dioxide would cause sea levels to rise by between 80 and 120 inches. By 1990 these estimates had been reduced by 75%. In 1996 a United Nations science advisory panel predicted a rise of only 15 to 22 inches by 2100—still based on shaky assumptions.

MEASURING CHANGES IN SEA LEVELS

Even these much smaller estimates are cause for skepticism, because changes in sea level are notoriously difficult to measure, and reliable information is hard to come by. All historical data are based on tide gauges, mainly on the shorelines of Northern Europe and North America. Long-term sea level trends must be extracted after adjusting for waves, storm surges and tidal variations. Scientists must also contend with the fact that the land surface may be rising or falling. Some land surfaces, in Scandinavia for example, are rebounding after being compressed by the weight of glaciers. Other land areas are subsiding, as petroleum or groundwater is pumped out.

Nevertheless, climate specialists have constructed a corrected "global" sea-level record; it shows that sea levels have been rising at the rate of about seven inches per century. Is this rise connected to climate? Not likely. The best estimates based on geologic data indicate that this has been going on for several centuries—a period in which the global climate has fluctuated significantly. Rather, the current thinking is that the steady rise in sea levels is connected to slow tectonic changes in the shape of the ocean basin—which human beings, of course, can do nothing about. Calculations of any climate effect on sea levels must take account of this more or less steady rise.

The next question is how global warming, if it occurs, would affect sea levels. On the one hand, a warmer climate would melt mountain glaciers and cause a thermal expansion of ocean water, accelerating sea levels' rise. But on the other hand, more water would evaporate from the surface of warmer oceans, leading to more rainfall, and—over Greenland and the Antarctic—to greater accumulation of snow and ice. This process essentially thickens the polar ice caps, thus lowering sea levels.

SEA LEVELS AND WARMING

While there is some evidence that sea levels have risen eighteen centimeters over the past one hundred years (with an uncertainty range of ten to twenty-five centimeters), there is little evidence that the rate of sea-level rise has actually increased during the time that, theoretically, warming has been accelerating. Says the International Panel on Climate Change (IPCC), "The current estimates of changes in surface water and ground water storage are very uncertain and speculative. There is no compelling recent evidence to alter the conclusion of IPCC (1990) that the most likely net contribution during the past one hundred years has been near zero or perhaps slightly positive."

Jerry Taylor, *Regulation*, Winter 1998.

The problem has been to find whether the sea levels' rising or lowering would have a greater impact. This is a question that cannot be answered by theory, or by computer models. To produce an accurate conclusion, we have to examine data. Fortunately, we do have data from a relatively rapid global warming that occurred between about 1900 and 1940, as the climate recovered from a cool period called the Little Ice Age. Neither the Little Ice Age nor the warming that followed are ascribed to human influences; many scientists believe they were caused by subtle changes in the sun's radiation.

In any case, data from the warming of 1900-1940 show a drop in sea levels, while the subsequent cooler period shows a sea-level rise. This effect is even more pronounced in comparisons of sea-level changes with sea-surface temperatures in the tropics, where most of the oceans' evaporation occurs.

ICE ACCUMULATION

These findings support the hypothesis that ice accumulation in the polar regions may have a greater impact on sea levels than do the melting of glaciers and the thermal expansion of ocean water. Support for this view also comes from concurrent but as yet incomplete measurements of ice accumulation at certain locations in Greenland and the Antarctic.

Contrary to activists' claims, what's clear is that global warming—if it takes place—would slow any rise in sea levels. Although more study is needed, this striking result should at least give pause to the Association of Small Island States and other groups that are pressing industrialized nations to adopt drastic reductions in fossil fuel use in order to avoid the global warming "catastrophe" so dear to the environmentalist lobby.

PERIODICAL BIBLIOGRAPHY

The following articles have been selected to supplement the diverse views presented in this chapter. Addresses are provided for periodicals not indexed in the *Readers' Guide to Periodical Literature*, the *Alternative Press Index*, the *Social Sciences Index*, or the *Index to Legal Periodicals and Books*.

Sophie Boukhari — "Marine Blues," UNESCO Courier, July/August 1998.

Merry Camhi — "Overfishing Threatens Sea's Bounty," *Forum for Applied Research and Public Policy*, Summer 1996.

Edward Carr — "A Second Fall," *Economist*, May 23, 1998.

Chris Chivers — "Troubled Waters," *E Magazine*, May/June 1996.

Jacques Cousteau, interviewed by Jim Motavalli and Susan Elan — "Jacques Yves Cousteau at 85," *E Magazine*, March/April 1996.

Ted Danson, interviewed by Campbell Wood — "Acting for the Oceans," *E Magazine*, January/February 1998.

Economist — "Too Many People Want Too Much from the Coast," May 23, 1998.

Don Hinrichsen — "Coasts in Crisis," *Issues in Science and Technology*, Summer 1996.

Bill McKibben — "The Earth Does a Slow Burn," *New York Times*, May 3, 1997.

E.G. Nisbet and C.M.R. Fowler — "Is Metal Disposal Toxic to Deep Oceans?" *Nature*, June 29, 1995. Available from Nature America, Inc., 345 Park Ave. South, New York, NY 10010-1707.

Patrick Nunn — "Global Warming," *New Internationalist*, June 1997.

John Prescott — "Seven Threats to the Seven Seas," *Our Planet*, vol. 9, no. 5, 1998. Available from the United Nations Environment Programme (UNEP), PO Box 30552, Nairobi, Kenya.

Carl Safina — "Scorched-Earth Fishing," *Issues in Science and Technology*, Spring 1998.

Mark Van Putten — "International Year of the Ocean," *International Wildlife*, July/August 1998.

CHAPTER 2

WHAT OCEAN MANAGEMENT AND CONSERVATION PRACTICES ARE BEST?

CHAPTER PREFACE

A fundamental question facing the community of nations is who possesses the oceans. While the earth's land has long been divided into nation-states with defined boundaries, the oceans have traditionally been viewed as a global commons owned by no one person or nation. Under traditions of international law dating back three hundred years, coastal nations exercised control of ocean waters that extended to three nautical miles from land. All nations in peacetime possessed equal rights to use the rest of the ocean for travel, fishing, transporting of goods, or scientific research without interference from other countries.

The question of who owns the oceans became increasingly important in the twentieth century as countries became more aware of the seas' potential riches, including minerals on the ocean floor. After World War II, a growing number of nations made claims of sovereign control over adjoining areas of the ocean beyond the customary three-mile limit. In an effort to resolve conflicts and create a uniform code of laws for the oceans, a series of conventions of the United Nations met to negotiate a system of rules governing boundaries, navigation, pollution control, and other maritime issues. Their efforts culminated in the Law of the Sea Treaty, a body of laws completed in 1982.

Among the features of the treaty was the creation of Exclusive Economic Zones (EEZs), which extended two hundred miles from land. In these areas, coastal nations had exclusive rights over fish, offshore oil, and other resources and had jurisdiction over research and environmental protection. In all, about 42 percent of the world's oceans are divided into EEZs. The rest of the oceans remain international waters open to all countries.

The Law of the Sea went into effect in 1994 after it was ratified by sixty nations. Controversy over the agreement and its implementation remains, however, and as of October 1998 it had yet to be ratified by the U.S. Senate (the United States did unilaterally assert its claim on its own EEZ in 1983). Supporters of the treaty argue that it brings to fruition the UN principle (declared in 1967) that the ocean's resources be made "a common heritage of mankind." Critics of the treaty argue that it is based on outdated socialistic principles in its mandates for the sharing of ocean resources between developed and undeveloped nations. The viewpoints in this chapter examine the Law of the Sea and other ocean issues facing the international community.

"It has been international law that
has provided the instrumentality
through which new . . . institutions
for ocean management have
emerged."

INTERNATIONAL REGULATIONS ARE NECESSARY TO PROTECT THE WORLD'S OCEANS

Lawrence Juda

Lawrence Juda is a political science professor at the University of Rhode Island and author of *International Law and Ocean Use Management*, from which the following viewpoint is excerpted. He argues that while international law has until recently treated the oceans as common property free for all to use with few restrictions, such a relatively anarchical "freedom of the seas" system is no longer tenable. Oceans, he asserts, require increasing legal and regulatory protection because of humanity's growing pressure on sea environments and resources. Expanded rules governing the use of the world's oceans must be international in scope, Juda maintains, because pollution, overfishing, and other ocean-related issues transcend national boundaries.

As you read, consider the following questions:

1. What assumptions about ocean resources have been made in the past, according to Juda?
2. How has technology affected international laws governing the use of the oceans, according to the author?
3. What are some of the accomplishments of the Third United Nations Conference on the Law of the Sea, in the author's opinion?

The use of the world's oceans raises questions which are central to the human experience. The oceans blanket some 71 percent of the earth's surface, provide the human race with food and recreational opportunities, serve as a highway for world commerce, and cover immense sources of usable energy and other nonliving resources. They have also been used inadvertently and purposefully as a sink in which to deposit the waste products of civilization.

Because of the pressure of growing population, particularly in coastal areas, and contemporary technology and associated effects, the human race is increasingly in a position to affect the workings of the oceans' natural systems, to deplete its renewable resources, and to detract from the natural beauty of the sea. Yet we also know more about the physical environment than did earlier generations and we continue to expand our understanding of the world around us. How do we apply all that we have learned over the years of the characteristics and natural processes of the oceans and the oceans' interplay with the earth's land masses?

In our treatment of ocean space we are displaying our approach to the wider question of human interaction with the physical environment. Myriad questions of values arise and we are forced to contemplate ever-widening systems of human cooperation to protect our common welfare. The need for legal and institutional mechanisms to manage and to regulate human behavior in relation to the world's oceans is ever more apparent.

THE NEED FOR OCEAN MANAGEMENT

The management of the uses of ocean space is a central focus of the field of marine affairs. Such management seeks, in accordance with some system of politically determined values, which is either explicit or implicit, to increase the benefits that may be derived from the resource and non-resource uses of the ocean. At the same time, it attempts to minimize detrimental effects on the ocean environment and to ameliorate conflict of use situations. In general it tries to provide for a directed balance among the various uses of ocean space as well as to protect the ocean environment from damage to its long-term viability.

But why and how did the need for such management develop? After all, for the most part, from the seventeenth century until very recently, the resources and uses of the seas beyond a narrow band of waters adjacent to the coast were available to all. In this earlier period the status of ocean space as a "commons," available to all but owned by none, seemed acceptable. Fish seemed to be limitless in the ocean and serious conflict of use situations were barely starting to emerge. In a world of very limited technological capability, with a relatively small population which typically lived out its life in a narrowly circumscribed geographic area around the point of birth, pressure on ocean resources and space was quite limited.

In such a world freedom of the seas as the organizing, legal principle could be seen as functional; however, freedom of the seas has proven increasingly dysfunctional in the face of the emergence of new technologies, a changing pattern of human use of ocean space, and altered human perceptions and expectations of ocean areas. Over time it became apparent that the various uses of ocean resources and space required some order. As a consequence of the status of most of the world's ocean space as a commons and because of the physical and biological characteristics of the oceans such that events and uses occurring in one place have impact in other locations, management efforts must of necessity involve some degree of international cooperation. . . .

A cursory consideration of daily world and national events may well suggest that the human race has many difficulties in managing terrestrial affairs despite thousands of years of experience. Given this history, what may we expect in attempts to manage the uses of the world's oceans? Yet despite any skepticism and reservations, there would appear to be no alternative to imposing some type of order in regard to the human use of the ocean environment and its resources.

REASONS FOR INCREASED REGULATION

The imperatives which drive us to this conclusion are varied and have a substantial cumulative weight. They include the following considerations:

1. *At least some of the important resources of the world's oceans are finite in human terms and the need for conservation of the limited resources found there is increasingly accepted.* The oceans, though vast, are limited in the resources of the greatest importance to mankind. What this means is that there are resources which may be exhausted through concentrated human effort. The potential to exhaust available resources is a relatively new factor and stems from at least three elements: growing demand pressures fed by sharply increasing populations with their attendant needs, the increase in expectations in lifestyles, which appears to be linked to greater levels of consumption of all types of material things, and the development of new and ever more effective means of exploitation of ocean resources.

Two important examples of such resources are ocean fisheries and offshore oil. Fisheries are of great importance since they provide sustenance to much of the world's population and for many serve as the main source of protein, without which proper human development does not appear possible. There is growing concern that the limits of exploitability have been reached in regard to a number of particular species of marine fish. Oil, a major component of the energy equation, particularly for the developed world, is increasingly being derived from offshore sources. Like its land-based counterpart offshore oil is finite in quantity. When it is gone, there is no more. In both instances, fisheries and offshore oil, it is important to maximize the rational exploitation and conservation of these resources so as to minimize waste. This requires some type of management effort.

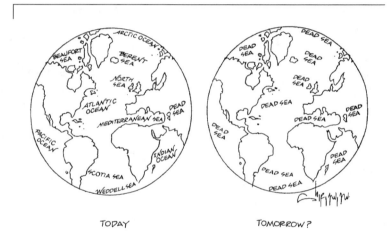

TODAY   TOMORROW?

2. *While the oceans cover some 71 percent of the earth's surface, the human use of the ocean is extremely concentrated, with by far the majority of activities occurring in areas close to the coast.* Two major reasons contribute to this pattern of human use. First, one must consider the simple concepts of propinquity and utility. Other things being equal, what is closer at hand gets used first. Why travel farther than is necessary to achieve the same result? Greater distances require greater efforts and involve greater costs. Whether in terms of fishing, waste disposal, or recreation it is simply easier to use resources which are nearby.

Second, while oceans extend over vast areas, the resources

which they contain or cover are not evenly distributed. The connection between the presence of fisheries and continental shelf areas of the world has long been noted. According to the Food and Agriculture Organization some 95 percent of the world fish catch comes from areas within 200 miles of the coast. Why then are these areas so heavily fished relative to areas beyond? Because of propinquity but also, in a variation of the famous statement of bank robber Willie Sutton, these areas are heavily fished because that is where the fish are. A similar situation is found in relation to offshore oil and gas. Deposits of such minerals are found in the continental shelf areas and, apparently, not beyond. Because of such realities, human activity in ocean areas is concentrated in but a portion of the world's oceans.

OCEAN USE CONFLICTS

3. *Conflict of use situations are becoming more apparent with attendant and associated opportunity costs.* As existing uses such as fishing intensify and diversify and as new uses emerge, the interference between and among uses increases. Vessel traffic in narrow waterways has grown. The amount of effort devoted to fishing activities has increased and new gear types and catch strategies have emerged which may have interactive effects on other uses or even other types of fishing. Some uses simply obviate the use of an area of ocean for other purposes. The dumping of toxic wastes, for example, makes the dumping area unusable for fisheries.

In general, it is now recognized that particular uses generate externalities: that is, unintended effects. Different uses of the same ocean space may coexist but they may require some degree of regulation to govern their relationship. Sea lanes and traffic separation schemes may be necessitated by the existence of offshore oil rigs. Without any type of regulation chaos could emerge with a variety of unpalatable ecological, economic, and political consequences.

4. *The protection of the marine environment has become an important value.* While human concern might at first focus on protection of resources it is becoming increasingly clear that attention must be given to the protection of the environment which sustains renewable resources and allows the enjoyment of the other uses of the sea. Some very special environmental systems ranging from coral reefs, to mangrove swamps, to the Antarctic may be especially vulnerable to environmental damage and require special attention.

In the broadest sense, the oceans sustain human life through their role in sea-atmosphere interchanges and in temperature

moderation. Increasingly, science is developing a better understanding of ecological interrelationships and the general web of
life systems. The human impact on the physical environment is
becoming more obvious over time, may threaten the normal cycles of nature, and may have untold consequences for human
survival and well-being.

5. *Human use of the marine environment and its resources has contributed to*
the potential for both international conflict and cooperation. As the significance of ocean resources and uses becomes ever more apparent,
given considerations of limited resources, the need for conservation and environmental protection, and the conflicts among
ocean uses, it is not surprising that international political and legal problems have arisen. Politically, who should benefit from
the uses of ocean resources and the ocean environment? Historically, in this century alone, the world has witnessed grave crises,
armed confrontations, and even wars relating to claims of freedom of the seas, fishing rights, and offshore jurisdictional
claims. . . .

New Priorities

On the threshold of the twenty-first century it is evident that the
pattern of ocean use as compared to that at the time of Hugo
Grotius [a Dutch jurist 1583–1645 whose writings helped establish international law] has changed radically both qualita-
tively and quantitatively. Spurred by the demands of a much
larger world population and a technological capability that
Grotius could not have even imagined, ocean uses have intensi-
fied and become more varied. Once perceived as inexhaustible,
fisheries in a growing number of regions have been exploited to
a level of commercial exhaustion and human impacts on the
physical environment in coastal areas have further threatened
the sustainability of living resources. The interaction of traditional uses such as fishing and new uses such as oil exploitation
have raised the specter of use conflicts. Though once seemingly
boundless, the oceans are increasingly being seen as systems
with limits. They cannot simultaneously afford the opportunity
for everyone to do as he wishes, without regard for the actions
of anyone else and without the growing risk of mutual interference, and the destruction of natural systems and the resources
those systems support.

It has become clear in this context that choices and priorities
have to be made because of the now widespread recognition
that some resources, such as desired species of fish, are finite in
quantity and that current actions may foreclose both present and

future options. The need for resource and environmental management within states is generally understood and institutional machinery in the form of national governments has the authority to make needed choices and decisions. While there is no guarantee that such actions are wise or effective, a recognized mechanism with the capacity to make decisions and to enforce them is in place.

Since the time of Grotius, however, the world's oceans, with very limited exception, have been treated as a commons and thus were immune to the exclusive legal authority of any single state. In a world with a relatively small population and, by today's standards, rudimentary technological capabilities, such a situation, in general, might be functional. In such a context, the concept of freedom of the seas still might be tenable. But, as the human capability to exploit resources grew enormously, and as the ability to affect natural systems became more evident and pronounced, freedom of the seas and its anarchical implications came to be questioned ever more. . . .

RETHINKING TRADITIONAL APPROACHES

As at the national level, change comes about when those in positions of authority come to accept that traditional ways of doing things are no longer effective or when alternative approaches are recognized as better serving contemporary needs. Technological change has had substantial impact, influencing international legal arrangements for ocean space. In regard to fisheries, for example, technological advances which revolutionized capture fisheries since the time of Grotius have forced a reexamination of the concept of freedom of the seas so as to protect stocks against decimation. The evolving legal system has allowed for extension of coastal state control and benefited coastal states at the expense of distant-water fishing states. Whether coastal states, however, can or will control their own fishermen or will allow national overexploitation and stock depletion remains to be seen. . . .

Technology, changing human understanding, and human needs have resulted in recognition of the necessity for management of mankind's use of ocean space. Because of the status of most of the oceans as a commons and the interplay of state interests, it has been international law that has provided the instrumentality through which new norms, arrangements, and institutions for ocean use management have emerged.

Obviously, modifications in the international legal system take time, patience, and effort to achieve. The international law system is ponderous, and different segments of the world community may not share the same sense of urgency for action. Further, the costs and benefits of such action must be apportioned among a variety of states, states which are often suspicious of the motives of others. Allocation of wealth as well as the conservation of resources and natural systems are at stake and each generation has had to come to terms with the many trade-offs implicit in any type of management scheme. Moreover, despite the availability of a greater amount of data and increased understanding of natural systems, decisions today are perhaps more difficult, given the pressures on resources caused by the world's vast and still growing population. Working out new international management arrangements will not be easy but there does not seem to be much alternative if renewable resources and the marine environment are to be protected and conflict of use situations minimized.

THE THREE DIMENSIONS OF OCEAN SPACE

In important respects ocean use management has come a long way since the days of Hugo Grotius, and now takes into account, as it must, the full three-dimensional character of ocean space. Modern technology has forced recognition of the different uses of the entire vertical column of ocean space and the possible interplay of different uses. This is reflected in the contemporary law of the sea, for example, in the creation of regimes for special-purpose zones, such as for continental shelves and exclusive economic zones. In these zones the coastal state is recognized as having very important but limited rights in parts of the vertical column of ocean space, while recognition of significant rights for other states in that column is maintained. This new international law takes into consideration activities beneath the sea floor, on the surface of that floor, in the water column, on the water surface, and in the airspace above. Accordingly, it has acknowledged the increasingly complex pattern of ocean use and, through the correspondingly more intricate legal regime it establishes, seeks to accommodate in a practical manner the growing multiplicity of uses, as well as the interests of individual states and those of the wider international community.

As the pace of change in the world appears to accelerate, an urgent matter of concern is whether the human race has the capacity to make needed alterations to its political, economic, social, and legal arrangements in a timely fashion so as to protect the human environment and the limited resources on which it depends for its survival. Clearly, for example, fisheries can be decimated much more quickly than was the case in the past and the capability to damage natural systems through human actions is much more potent than it was in earlier times. . . .

A limitless conception of freedom of the seas increasingly proved unworkable over time. Attempts to alter governing international law met substantial resistance but change has occurred, nevertheless, and is apparent to those who compare the ocean law system of the Grotian period to that of the contemporary world. The most recent attempt to codify general ocean law, the Third United Nations Conference on the Law of the Sea, marks a major event in the ongoing development of that law, expanding national authority in offshore areas and, just as importantly, establishing the legal obligations of states for the responsible management and protection of marine resources and the environment. The convention it drafted represents a significant retreat from a laissez-faire system of free and open use of ocean space beyond narrow territorial seas and movement toward a system of regulated and responsible use which considers tomorrow's as well as today's needs. Further, it provides a framework for future development and establishes important mechanisms for dispute settlement, the decisions of which may make further significant contributions to the evolution of internationally accepted ocean management systems.

"There seems to be a gulf between
what governments are committing
themselves to do in international
law and what they are doing."

INTERNATIONAL REGULATIONS HAVE FAILED TO PROTECT THE WORLD'S OCEANS

Peter H. Pearse

The following viewpoint is taken from a presentation made by Peter H. Pearse at an international academic conference on the oceans and international law. Pearse questions whether the numerous international treaties and regulations designed to protect the oceans are working effectively. While most coastal states now have laws and international obligations directing them to conserve fish resources and limit pollution, Pearse argues that many countries are failing to implement sound conservation and sustainable development practices. Rather than create new laws and regulations, he asserts, countries should attempt to implement existing regulations more effectively with the use of market incentives. Pearse is a professor emeritus of forestry at the University of British Columbia in Vancouver, Canada.

As you read, consider the following questions:

1. What key question does the author attempt to answer about international laws and regulations on ocean management and conservation?
2. What benchmark does Pearse use to evaluate ocean management policies?

Developments in marine environment protection are taking place not only rapidly but also on a wide front. Most important is the entry into force in November 1994 of the United Nations Convention on the Law of the Sea (UNCLOS, Part XII of which is particularly relevant to protection of the marine environment) and Agenda 21 of the 1992 United Nations Conference on Environment and Development (UNCED). Coastal states around the world are also busy independently adopting measures to protect ocean environments under their jurisdiction. Also involved are regional bodies, such as the European Union. . . . And at the global level, a host of international organizations, conventions, and treaties bear on protection of ocean environments. . . .

KEEPING PACE WITH ENVIRONMENTAL PRESSURE

The question is whether all the recent international regulatory effort is achieving its objective of protecting ocean resources to the degree that policy-makers seek. Or is it too meager or too slow to deal with the problem, allowing ocean resources to deteriorate, to the prejudice of sustainable development?

In historical perspective, the recent development of regulatory instruments and international law for protecting marine environments has certainly been massive and rapid. But threats to ocean environments have also been increasing and proliferating. Since mid-century we have witnessed widespread pollution of coastal waters, massive growth in fishing pressure and overfishing of stocks, emergence of a vast new array of toxic and hazardous wastes, substantial increases in the scale of shipping and other methods of transporting potentially dangerous goods, the spread of waste and persistent chemicals, and a host of other well-known hazards to the marine environment.

The question, then, is whether the rapid development of regulatory instruments is adequate to control the rapid development of threats to the ocean environment. Undoubtedly, progress has been made. Some observers claim that, generally, we are winning the war against environmental degradation, pointing to substantial reductions in pollution, improvements in air and water quality, and expansion of forests in industrial countries. More pessimistic commentators argue that institutions change too slowly to keep pace with today's accelerating changes in economic activity, technology, and public demands, portending the "future shock" described by American futurist Alvin Toffler.

Whatever the success in protecting land and atmospheric environments, it is difficult to find much evidence of improvement in the marine environment over the last half century.

Notwithstanding the impressive efforts of national governments and international organizations, threats to the marine ecology are greater than ever before and continue to grow, pollution of most kinds continues to increase, and more and more of the oceans' living resources are being depleted.

SUSTAINABLE DEVELOPMENT

To evaluate progress, we need a frame of reference; that is, what our policies are aimed at achieving. The idea of sustainable development, now widely accepted in principle, provides a useful starting point. . . .

Notwithstanding ambiguities about the operational meaning of sustainable development, it is sufficiently clear about its two main tenets. One is its environmental principle: the integrity of ecological systems must be protected to maintain their biological productivity and diversity. The other is its development principle: resources must be managed and used in ways that will accommodate increasing economic productivity over time. These goals are seen to be related and complementary.

The two most conspicuous failures in the pursuit of sustainable development of the oceans are the failure to protect marine environments from pollution and the failure to prevent overexploitation and waste in ocean fisheries. The problem of pollution has been the subject of other papers, which describe continuing deterioration of marine environments, especially from land-based sources. (Apparently some success has been achieved in reducing pollution from ships, but this is a relatively minor part of the problem of marine pollution.) Conservation of fish has been a subject of national and international regulation longer than any other aspect of the marine environment, yet the general condition of the world's fisheries is characterized by overexploited and depleted resources and waste of natural and human resources. Decades of regulatory effort have left the fisheries among the most highly regulated economic activities, yet the condition of the stocks almost everywhere is bleak and worsening.

THE FISHING CRISIS

Through most of the second half of this century, the world catch of fish continued to increase, but this was not so much a result of improved management and utilization of stocks as a process of moving on from depleted stocks to new stocks. Today, the frontier of economic exploitation is closed. The world catch has peaked at nearly 100 million tons per year, but at least another 30 million tons is caught accidentally, discarded, and wasted. Most of the world's most valuable stocks are overfished and declining.

A U.N. Food and Agricultural Organization (FAO) report indicates that 70 percent of the world's marine fish stocks are heavily exploited and depressed. Of the FAO's fifteen major fishing areas, catches have declined in all but two. Declining catches are reported in all regions of the Atlantic, all regions of the Pacific, and in the Mediterranean and Black Seas. In many important fisheries, such as Atlantic cod and Peruvian anchovy, catches are now a small fraction of past levels.

The world's ocean fisheries show a conspicuous failure to improve economic productivity as well. Vastly overexpanded fishing fleets, with fishing power far in excess of that needed to take the sustainable catch, not only frustrate efforts to conserve the stocks but also involve enormous waste of capital and labor. Unlike almost all other natural resource industries, which show increasing productivity, the economic performance of fishing industries continues to decline. This is the antithesis of sustainable development.

THE GULF BETWEEN COMMITMENT AND PERFORMANCE

As a non-expert observer, I am struck by two stark facts. One is that the new law of the sea obliges coastal states to prevent pollution and to conserve fish resources in waters under their jurisdiction. I know there are many qualifications and ambiguities about these obligations, having to do with economic exigencies, straddling stocks, and other matters. But notwithstanding these qualifications, the basic proposition seems clear: coastal states must take all necessary measures to prevent pollution and prevent depletion of fish stocks in their areas of jurisdiction.

The other observation is that coastal states are failing to meet these obligations on a massive scale, apparently with impunity. To this extent—and it is a very large extent—the new law of the sea is not working. Are international law-makers whistling in the wind?

The failure, it seems to me, is not due to insufficient law—at least not insufficient international law. The failure is a failure to implement the law. There seems to be a gulf between what governments are committing themselves to do in international law and what they are doing.

In international circles, lawyers and diplomats are busily making commitments to fashionable new concepts like ecosystem management, sustainable development, integrated resource systems, and precautionary principles. Meanwhile, their regulatory agencies are wrestling with the practical problems of controlling pollution and managing fishing, with limited success. There is a gap between the promise and the practice. And it is a widening gap.

LIMITATIONS OF REGULATORY POLICIES

We must ask why governments have apparently found it so difficult to implement these laws. I would like to suggest that they may be taking the wrong approach. Hitherto, the approach taken by most states in pollution control and fish conservation has been almost entirely regulatory; that is, it depends mainly on enacting restricting laws, regulations, and controls that are designed to stop polluters and fishermen from doing things that they want to do, such as to get rid of their wastes as cheaply as possible, and to catch as many fish as possible.

This is the traditional regulatory approach: when private producers or consumers behave in ways that threaten the public interest, governments enact regulations to constrain them, subject to penalties, and then take measures to police behavior and enforce the rules. But enforcing rules that run counter to the economic interests of thousands of users of the marine environment is increasingly costly and difficult. And it is proving to be too costly and difficult, and too unpopular among the vested interests, for most governments to do successfully.

THE ALTERNATIVE OF ECONOMIC INSTRUMENTS

An alternative approach is to cultivate economic incentives for producers and customers to behave in the desired way—to redirect market forces away from behavior that offends the environment. This involves developing economic instruments that encourage people who use resources to do so in ways that are consistent with policy objectives and to encourage them to take an interest not only in using the resources but also in protecting them, managing them, and enhancing them. The purpose is to make market forces work better, so regulatory intervention is less necessary.

The potential for economic instruments in environmental management has attracted considerable attention in economic literature recently, and they have been developed and implemented to deal with a growing variety of environmental problems. These instruments take a variety of forms: charges for resources that are traditionally available free (such as water) to discourage waste and encourage conservation; fees for disposal of wastes into the atmosphere, waterways, and on land to encourage polluters to abate their wastes; and (most relevant in

GOVERNMENT REGULATION DOES NOT WORK

The ocean is an economic "commons." Single individuals and firms do not have exclusive claim to portions of the sea's bounty. There is every incentive for any one fisherman to catch as many fish as possible. After all, if he does not, someone else will. If there is one fish left, it is to the individual fisherman's advantage to catch it. There is no incentive or advantage for him to invest in enhancing ocean productivity, since others who do not bear the costs of fertilizing the oceans will reap the rewards. Both aspects of the "commons" problem must be solved in order to enhance ocean resources. . . .

One method of dealing with the problem is to use government regulation. That has been tried in New England and many other fisheries with uniformly poor results. The method of regulation used often involves limiting access to fishing; limiting the technology applied to catching the fish; and limiting the pounds of fish that may be caught. It is always to the fisherman's advantage to ignore or circumvent the regulations, since he gets no return for fish left in the sea. Also, government regulators respond to political pressures and have no stake in maximizing the output of a resource. . . .

The most effective approach would . . . be to privatize the entire resource.

Michael Markels Jr., *Regulation*, vol. 18, no. 3, 1995.

this context where users' rights to resources are commonly poorly defined and held in common) the development of well-defined property rights, to give users incentives to protect and conserve the resources and thereby the value of their rights.

Property rights have been developed in increasing variety in recent years to support the management of water flows, water quality, air emissions, and fish and wildlife with considerable success. Among the best known are the tradable emissions permits issued to coal-burning utilities in the United States. In fisheries, perhaps the most promising development in recent years has been the development of property rights to shares in the catch, issued to individual fishermen, which are divisible and transferable. This approach is best developed in New Zealand, where it has been highly successful in improving both the management and conservation of fish stocks and the economic performance of the fishing industry. Instead of always trying to catch more fish, the holders of fishing rights take a strong interest in protection and conservation of the resources on which their valuable rights depend, and sometimes press the government for even more conservative policies. While this is not the place to describe the variety and form of economic instruments, it is important to note that recent experiments show considerable promise.

TURNING AWAY FROM GOVERNMENTS

In this context, it is important to note, also, that while we have been loading massive new environmental responsibilities onto governments, most of the world is turning away from reliance on governments to solve problems. This is conspicuously so in the former socialist countries of the eastern bloc, but it is also true in the capitalist west, where most governments are currently involved in programs of privatization, deregulation, and government downsizing.

Moreover, attitudes toward economic organization and governance around the world are converging in this receptiveness to market mechanisms. An implication of the recent efforts to develop market economies in countries of the former Soviet Union, Eastern Europe, and China is that there are no longer many governments whose economic doctrine denies the value of natural resources, rejects private property, and otherwise impedes the use of economic instruments.

My comments thus converge with the conclusions of Edward Miles in his presentation at the beginning of this conference:

> . . . the need is not for more law. The critical need instead is for drastic improvements in institutional design, policy design and policy implementation.

If our frustrations in trying to protect the marine environment are not due to insufficient laws but inadequate implementation, policymakers and lawyers should pay more attention to the problems of implementation. Specifically, we should be searching for ways to harness economic incentives to induce producers to shift their behavior away from activities that harm the environment and toward behavior that protects and enhances it.

As we try to cope with growing pressures on marine resources and ocean environments, there is a danger that we will become locked onto a treadmill of governmental regulation. This is not to say that we will need no more laws, and there will always be a place for regulatory controls. But to make progress in this area we may have to lighten the burden we are putting on traditional restrictive regulation and instead harness market forces whenever we can.

Sustainable development means harmonizing economic activity

with the environment. The driving force of economic activity is the incentives of producers and consumers, and the power of these economic forces is formidable. Our approach to environmental problems has been to try to restrict these forces. But harmony between the economy and the environment might be better achieved by ensuring that market forces work in environmentally friendly ways.

"Conservation and control are the raisons d'être of the law of the sea."

THE LAW OF THE SEA CONVENTION SHOULD GOVERN THE USE OF OCEAN RESOURCES

Ian Townsend-Gault and Hasjim Djalal

The United Nations Convention on the Law of the Sea, a body of international law negotiated between 150 nations, entered into force on November 16, 1994, one year after it was ratified by sixty nations. In the following viewpoint, Ian Townsend-Gault and Hasjim Djalal describe the evolution of the Law of the Sea and other international laws concerning the ocean and examine some of the issues such laws are meant to address, including fishing rights, offshore oil drilling, environmental protection, and deep seabed mining. They argue that the Law of the Sea will prevent conflict on the oceans and ensure that marine resources will benefit all nations, including landlocked and less developed countries. Townsend-Gault is a law professor at the University of British Columbia in Vancouver, Canada. Djalal, an Indonesian diplomat who participated extensively in the Third United Nations Conference on the Law of the Sea, is director of the Centre for Southeast Asia Studies in Jakarta, Indonesia.

As you read, consider the following questions:

1. What motivated many countries to extend their control over coastal waters in recent decades, according to Townsend-Gault and Djalal?
2. What ethical problem do the authors say was created by the spread of national jurisdiction over the oceans?
3. How many nations have ratified the Law of the Sea at the time the viewpoint was written?

Reprinted from Ian Townsend-Gault and Hasjim Djalal, "Laws for the Ocean," UNESCO Courier, July/August 1998, with permission.

The largest treaty ever negotiated, the United Nations Convention on the Law of the Sea is a fairly recent development, entering into force on 16 November 1994. Yet its roots stem back to the early 1950s when the UN International Law Commission was set up with the task of codifying and developing international law, a great deal of which is unwritten, arising from the consistent practice of states. While this customary approach is excellent for establishing principles, treaties are needed to work out the essential details.

NEW AND OLD ISSUES

For more than 300 years, it was simply accepted that coastal states are each entitled to sovereignty over a body of water immediately beside the coast. This sovereignty was absolute except for the right of innocent passage by foreign merchant vessels. However, international law did not, for example, define the exact extent of this "territorial sea" (which was three and is now twelve nautical miles) or the conditions of "innocent passage". So in about 1951, the International Law Commission began working to resolve these old issues as well as some new ones: namely the conservation of the ocean's living resources and petroleum development on the continental shelf.

Concerned by dwindling fish stocks, countries like Portugal, Spain, Argentina, and Iceland had been pressing for decades for zones of fisheries conservation jurisdiction. However, they were outnumbered by the majority of coastal states who wanted to preserve more or less unlimited and unrestricted access to the living resources in what were still regarded as the high seas.

Turning to offshore petroleum, geologists had long known that the continental shelf had considerable potential. Declining domestic sources of oil coupled with increasing domestic demand led American experts to look closely at the seabed of the Gulf of Mexico beyond the three nautical mile limit. Yet there was a problem. While companies were not prohibited from exploring beyond that limit, they enjoyed no legal security in conducting operations there. They also had no "legal" rights to any petroleum they might find. In order to provide a proper legal foundation for offshore petroleum activities (and the steep investment required), the American government proposed that all coastal states should have the right to extend their jurisdiction to the edge of the "continental shelf".

This proposal was accepted with great enthusiasm throughout the world. However, the motives for agreement were not all the same. For some countries, the extension was a cloak for control over the adjacent fishery. This fundamental difference prompted the International Law Commission to step in and lead the way to the First United Nations Conference on the Law of the Sea, convened in Geneva in 1958. Yet despite the fact that four Conventions—on the Territorial Sea and Contiguous Zone, the High Seas, the Living Resources of the High Seas, and the Continental Shelf—were accepted and signed, there was an obvious divergence between countries in favour of fisheries jurisdiction and those opposed. As a result, there were major gaps in some of the treaties.

UNITED STATES SHOULD SUPPORT THE LAW OF THE SEA

The Law of the Sea Convention, as now modified, provides the comprehensive, international framework that our country has sought for many years. Apart from the recently concluded trade agreement, it is perhaps the most significant example of the attainment of consensus in a major multilateral negotiation since the establishment of the United Nations—a consensus which in all respects serves to buttress U.S. national security needs while providing important tools to protect our marine environment and promote the sustainable use of ocean resources. It would be a great achievement in international law for the United States to join with other major maritime powers and now formally consent to accession to this convention.

David A. Colson, *U.S. Department of State Dispatch*, February 13, 1995.

A second conference convened in 1960 to resolve the issue was an absolute failure. From then on, states which considered that they had been victimized by the lack of jurisdiction felt obliged to take the law into their own hands. Countries like Iceland started enforcing exclusive fisheries zones to 12, then 50, then 200 nautical miles. Others followed suit; some willingly, others less so, but the former eventually prevailed over the latter while the Convention was still being negotiated.

Meanwhile, progress was made in dealing with mineral resources. Countries looked favourably upon the Convention on the Continental Shelf of 1958 which recognized "sovereign rights for the purposes of exploration and exploitation" on the seabed and subsoil of the shelf. It is important to note that the terms "sovereignty" or "ownership" are carefully avoided. The Convention's framers did not want to suggest that countries exercised absolute rights beyond the limits of the territorial sea, because this might ultimately endanger the freedom of the seas.

RIGHTS AND FREEDOMS

The problem lay in defining the extent of these rights. A fixed limit was seen as artificial. On the other hand, a reference to the geographical feature "continental shelf" would not please coastal states like Chile which does not have a natural shelf of significant extent. So the edge of the "legal" continental shelf was fixed at the 200-metre isobath (the world's average shelf edge depth), or beyond "to the point where the depth of the superjacent waters admits of exploitation". In other words, all countries were guaranteed a minimum of control to a 200-metre depth, while those that had the technology to exploit deeper could still do so. We must remember that in 1958, 200 metres seemed to be an extraordinary depth. The problem was, of course, that science and technology developed faster than anyone had thought possible.

So by the early 1970s, it looked as if industrialized coastal states were about to carve up the world's ocean space. This was unacceptable for a number of reasons. First, it looked like a bonanza for coastal states at the expense of their land-locked cousins. Second, accidents of political geography sometimes resulted in curious boundaries. Countries with relatively long coastlines were clearly going to do better than others. Portugal, or Chile with its 4,200 kilometres of coast would have a relatively vast area of ocean space compared, say, to Spain.

There was also a serious ethical problem. If national jurisdiction was allowed to spread unchecked, it would be virtually impossible to fairly distribute the oceans' resources. To counter this, attention turned to limiting this jurisdiction and recognizing the areas beyond as the "common heritage of mankind".

A THIRD INTERNATIONAL CONFERENCE

Given the dimensions of the problems, the time was clearly ripe for another international conference. For example, the archipelagic states wanted recognition of their rights to the waters between and around their islands. Landlocked states wished to have their rights of access to marine resources clearly articulated. And there was an urgent need for detailed rules regarding the preservation of the marine environment—a topic virtually ignored in 1958. And who was to govern and profit from the area beyond national jurisdiction? There was also some feeling that the International Court of Justice could not properly resolve these issues. A tribunal was needed to specialize in the Law of the Sea.

A third UN conference ended in December 1982. All the participating countries voted for the Convention except four—Israel, Turkey, the United States, and Venezuela—and seventeen that abstained. Until recently, the countries becoming full parties to the Convention were found in Asia, Africa, and South America. This, and the continuing refusal of the United States even to sign the Convention, led some observers to dismiss the conference results, but it is now clear that such judgments were premature.

MINING THE SEABED

The major difficulty lies with mining the deep seabed, which is the area considered to be the common heritage of mankind. US firms prevailed on the Reagan administration not to sign, much less ratify, the Convention. However, a compromise on the regime for such mining was reached in 1994 which should prove acceptable to all (and is to the industrialized countries).

Yet as in 1958, not everything is perfect. For example, Canada's east coast has one of the world's most extensive continental shelves. According to the Convention, petroleum activities in this area are firmly under Canadian jurisdiction. However, control over living resources extends only to 200 nautical miles through the creation in 1994 of the Exclusive Economic Zones. The problem is, Canada's shelf goes well beyond this distance, with species swimming in and out of the zone. So, in theory, foreign vessels can remain just outside Canadian jurisdiction and catch as many fish as they please, to the possible detriment of Canadian fishers who are obliged to follow Canadian rules in or out of the zone. This has led to both a confrontation and treaty between Canada and Spain; the problem was resolved by an Agreement on High Seas Fisheries in 1995.

CONSERVATION AND CONTROL

Conservation and control are the raisons d'être of the law of the sea. The concept of the "common heritage of mankind" recognizes that a country should not have the right to exploit a resource (e.g., the deep seabed) merely because it has the capacity to do so. The common heritage is held in trust for the peoples of all countries. The Convention forbids exploration or exploitation there without a mandate from the International Seabed Authority (established by the Convention and based in Jamaica) which must ensure that the benefits accruing are distributed equitably. The other expert bodies are emerging: the Tribunal on the Law of the Sea was established in Hamburg in 1996, and the Continental Shelf Commission now meets in New York. In short, the process is well underway.

No one should underestimate the challenges involved in implementing the Convention, which has now been ratified by 125 states. Yet in light of the fragility of our ocean environment, it is in the interest of both the industrialized and less-industrialized states alike to overcome them.

"Created at a time when statism held
sway internationally, the Law of the
Sea remains a bad agreement."

THE LAW OF THE SEA CONVENTION SHOULD NOT GOVERN THE USE OF OCEAN RESOURCES

Doug Bandow

In 1982 President Ronald Reagan announced that the United
States would not become a signatory to the United Nations Convention on the Law of the Sea, a comprehensive collection of international laws that was the product of nine years of negotia-
tions within the UN. In 1994, after parts of the agreement
relating to seabed mining had been renegotiated, the United
States dropped its opposition to the treaty and President Bill
Clinton sent it to the Senate for ratification. In the following
viewpoint, Doug Bandow argues that the Senate should reject the
treaty. He maintains that despite its modifications, the treaty is
still based on discredited socialist premises and would unfairly
burden companies and countries that attempt to mine the
seabed. Bandow, an author and researcher affiliated with the Cato
Institute, served as a member of the U.S. delegation to the Third
United Nations Law of the Sea Conference.

As you read, consider the following questions:
1. What was the central purpose behind the original Law of the
   Sea Treaty, according to Bandow?
2. What, in the author's opinion, is the practical impact of the
   Law of the Sea Treaty that will go into effect with or without
   U.S. compliance?

Abridged from Doug Bandow, "Faulty Repairs: The Law of the Sea Treaty Is Still
Unacceptable," Foreign Policy Briefing No. 32, September 12, 1994. Reprinted by
permission of the Cato Institute. *Endnotes in the original have been omitted in this reprint.*

In Washington bad ideas never die. They simply lie dormant, waiting for a sympathetic bureaucrat or politician to revive them. So it has been with the Law of the Sea Treaty, which covers everything from navigation to seabed mining. Although rejected by President Ronald Reagan more than a decade ago, the LOST, as it is called, has returned to life under the Clinton administration. After winning a few changes in the treaty's most burdensome provisions, the State Department has now enthusiastically endorsed the agreement. On July 27, 1994, before the United Nations (UN) General Assembly, U.S. ambassador Madeleine Albright praised the LOST for providing "for the application of free market principles to the development of the deep seabed" and establishing "a lean institution that is both flexible and efficient." Two days later Albright formally affixed her signature to the convention, which now goes to the Senate for ratification. [The Senate had not ratified the treaty as of August 1998.]

Although the revised LOST is not as bad as its predecessor, it would still create a Rube Goldberg system—with the International Seabed Authority (ISA), the Enterprise, the Council, the Assembly, and more—that would be guaranteed to become yet another multilateral boondoggle. It would not only waste money but also discourage the production of ocean minerals. Moreover, the treaty would resurrect the redistributionist lobbying campaign once conducted by developing states unwilling to deal with the real causes of their economic failures. Indeed, the LOST would essentially create another UN agency with the purpose of transferring wealth from industrialized states to the Third World voting majority.

THE NEW INTERNATIONAL ECONOMIC ORDER

Throughout the 1970s and early 1980s, most Third World states saw socialism as the wave of the future. At home they implemented centrally planned development strategies. Abroad they promoted what they euphemistically called the New International Economic Order—global management and redistribution of resources, technology, trade, and wealth. Their UN lobbying arm, the so-called Group of 77, pressed for technology transfers, corporate codes, controls on transborder data flows, restrictions on investment, limits on intellectual property rights, international taxation, commercial preferences, additional foreign aid, and domination of unowned natural resources. Virtually no international organization was untouched by that campaign: such alphabet-soup agencies as the United Nations Industrial Development Organization (UNIDO), the

United Nations Educational, Scientific and Cultural Organization (UNESCO), the Food and Agriculture Organization of the United Nations (FAO), the United Nations Centre for Transnational Corporations (CTC), the United Nations Conference on Trade and Development (UNCTAD), and the United Nations itself became international battlegrounds.

No fight was more important than that over the LOST. The treaty promised to "contribute to the realization of a just and equitable international economic order which takes into account the interests and needs of mankind as a whole and, in particular, the special interests and needs of developing countries." It took a decade to negotiate and ended up containing more than 400 articles on subjects ranging from ocean transit to marine pollution to territorial seas. But its most hotly contended provisions addressed a then nonexistent industry: seabed mining.

THE COMMON HERITAGE OF MANKIND

The ocean floor is littered with manganese nodules and deposits of other resources, such as cobalt, coral, oil, and polymetallic sulfides. The prospect of recovering what was believed to be virtually untold wealth began to gain attention in the 1960s; in 1967 Malta's UN representative, Arvid Pardo, proposed that the seabed be declared the "common heritage of mankind." The UN General Assembly adopted Pardo's rhetoric, established the ad hoc Seabed Committee, and proposed creation of a system to ensure "equitable sharing by States in the benefits derived" from the seabed. In 1973 the United Nations organized the Third United Nations Conference on the Law of the Sea (the first two had focused on ocean jurisdiction and fishing). Eleven sessions later, on April 30, 1982, UNCLOS III brought forth a treaty that theoretically guaranteed the ocean's resources as the common heritage of mankind.

The convention went on to create the ISA and the Enterprise, to mine the ocean floor for the ISA. The ISA was to be ruled by the Assembly and the Council; the Soviet bloc was to be guaranteed three seats on the latter while the United States was assured of none. Western countries and firms were to provide funds and technology to the ISA for redistribution to developing states. The treaty limited production, failed to guarantee private firms nondiscriminatory access to the seabed, subsidized the Enterprise, and otherwise promoted economic war on the industrialized states. That monumental boondoggle was admittedly bipartisan in the making: Henry Kissinger, Elliot Richardson, Jimmy Carter, and Alexander Haig all supported it. But President Ronald Reagan did not, and he was soon to be followed by the leaders of other major developed states and even the Soviet Union in rejecting the accord. Proponents of the LOST warned that chaos and doom were sure to result. The world went on as before, however, and the agreement sank beneath the waves.

But diplomats are attracted to treaties like moths to lights. Informal discussions began in 1990 over the possibility of the United States' adhering to a revised agreement and culminated in President Bill Clinton's decision to sign the LOST and submit it to the Senate for ratification. Stated Secretary of State Warren Christopher, "It's an extremely important treaty, and I think it's very desirable that we have been able to obtain from the other members satisfactory amendments to the seabed mining provisions that enable us to approve the treaty as a whole."

A TREATY IN SEARCH OF A PROBLEM

The basic question, which the secretary did not answer, is, why the LOST? David A. Colson, deputy assistant secretary of state for oceans, admitted that "the basic flaws of that deep seabed mining regime are manifold." But the administration obviously assumed that the United States should sign, so it directed chief negotiator Wesley Scholz and his colleagues to "fix" the treaty. Is there any reason to join, however? Haggling over which of the accord's tentacles are most dangerous and therefore should be severed seems to have obscured the fact that the whole octopus should be killed.

Some advocates of the LOST have argued that a universal accord would promote seabed mining. A truly market-oriented accord might do so; not, however, a system that includes the ISA, the Enterprise, the Council, revenue sharing, international royalties, Western subsidies for the Enterprise, a Council veto for land-based minerals producers, and the like. Yet all of those anti-development provisions remain in the revised text.

That the treaty would favor political over productive activity should come as no surprise. The LOST was created in a different era. It was intended to inaugurate large and sustained wealth transfers from the industrialized states. The structure was therefore crafted to advance ideological, not economic, goals. Since then, however, most developing states have moved away from collectivism, and the promise of undersea mining has largely evaporated. Yet the original collectivist framework remains. Even the State Department acknowledges that the new "agreement retains the institutional outlines of Part XI," which contains the seabed mining provisions. The treaty has become a solution in search of a problem.

A good international treaty would be useful, but it is not necessary. True, Elliot Richardson, who led the American delegation during the Carter administration, claims that the United Nations' mere assertion that the ocean's resources are the "common heritage of mankind" has abrogated any right to mine the seabed without that body's approval. He warns that "if any mining defied international law, its output would be subject to confiscation as contraband." Ambassador Richardson does not explain who would do the seizing—a UN navy? More important, until Washington and its industrialized allies ratify the LOST, their nationals retain the liberty to mine the seabed. In fact, that makes it all the more important that the United States refuse to ratify the accord. Once Washington has done so, a future renunciation of the LOST might not be considered enough to reestablish Americans' traditional freedom on the high seas.

Still, treaty supporters point out, the LOST, having gained more than the necessary 60 ratifications from UN member nations, will go into effect in November 1994 irrespective of Washington's ratification decision. However, nations cannot be held to surrender their rights because other states have ratified a treaty. Put bluntly, it matters little whether or not Djibouti, Fiji,

---

POTENTIAL TRAPS

While the United Nations Convention on the Law of the Sea (UNCLOS) has effectively codified many aspects of traditional law and has successfully incorporated several modern issues, such as environment, fisheries, and coastal zone management, these can be regarded as "nice to have" accomplishments but are by no means essential to the political, economic, or military security of the United States. . . .

On the other hand, the regulatory, political, technological economic, and possibly military concessions embedded in the treaty represent a set of potential threats and traps that the United States should not walk blithely into. . . .

The International Seabed Authority and UNCLOS represent the surrender, with little or no compensation, of a variety of tangible U.S. security and sovereignty equities over a geographic area encompassing 70 percent of the Earth's surface. The administration is attempting to bind this nation to a treaty and a bureaucratic organization whose basic operating principles are inimical to U.S. interests and that, to date, is officially recognized only by third-world and landlocked states.

Peter M. Leitner, *Reforming the Law of the Sea Treaty*, 1996.

---

or Zambia approves of American mining consortia operating in the Pacific. An ISA without any industrialized states as members would be about as effective as the "international regime" that is supposed to be established under the UN Moon Treaty, which—I am not making this up, to quote humorist Dave Barry—formally took effect in 1984.

A decentralized and relatively informal system, perhaps with a small international office, that provided for mutual recognition of mine sites and arbitration of conflicts would offer adequate security of tenure for mining companies. In fact, the United States and the Europeans implemented that type of strategy when they rejected the LOST. Other nations, particularly those like China, India, and South Korea, which have indicated an interest in seabed mining, could be invited to join such a system as well. That approach would operate with minimal bureaucracy and cost and would be confined to essentials—most important, developing a stable investment regime.

Supporters of the LOST point out that the convention covers other subjects, such as navigation and scientific research, and that those provisions are generally noncontroversial. They are also largely irrelevant, because most merely codify customary international law, proving that peaceful cooperation and global order can develop without a new UN agreement and organization. As for the most serious questions of naval transit, few countries have the incentive or ability to interfere with American ships. Not once during the last decade has a U.S. vessel been denied freedom of transit. Anyway, it is the U.S. Navy, not the United Nations' LOST, that will ultimately guarantee American interests. If Part XI were acceptable, there would be little harm in acceding to the other provisions. But the nonseabed sections do not compensate for the flawed seabed mining regime.

INADEQUATE REPAIRS

True, the administration argues that it has transformed the treaty. "We have been successful in fixing all the major problems raised by the Reagan administration," explained Scholz. "We have converted the seabed part of the agreement into a market-based regime."

Well, not quite. Scholz and his colleagues did work hard to turn a disastrous accord into a merely bad one. The result is an improvement—and a dramatic testament to the distance that market ideas have traveled since the LOST was opened for signature in 1982. But the ISA remains an unnecessary boondoggle, intended only to hinder seabed development. The Enterprise continues to be an economic white elephant. The financial redistribution clauses remain a special-interest sop to poor states. And the entire system is likely to end up as bloated and politicized as the rest of the United Nations.

Moreover, for all their emphasis on individual problems, the negotiators have left a number of the worst ones unsolved. In places they have substituted ambiguity for clearly negative provisions. For instance, the treaty retains both the ISA, of undetermined size, and the Enterprise, an international version of the ubiquitous state enterprises that have failed so miserably all over the world. The ISA remains almost comically complicated, with its Assembly and Council and such subsidiary bodies as the Finance Committee and the Legal and Technical Commission, all with their own arcane rules for agendas, membership, procedures, and votes. The LOST revisions restrict some of the ISA's discretion but still submerge seabed mining in the bizarre political dynamics of international organizations. Private firms must continue to survey and provide, gratis, a site for the Enterprise for each one they wish to mine. Anti-monopoly and anti-density provisions still apply disproportionately to American mining firms.

ISA fees have been lowered, but companies will continue to owe a $250,000 application fee and some, as yet undetermined, level of royalties and profit sharing. . . .

An additional problem occurs because the land-based mineral producers, whose interest is antagonistic to the very idea of seabed mining, and "developing States Parties, representing special interests," such as "geographically disadvantaged" nations, each have their own chamber and thus a de facto veto over the ISA's operations. Moreover, the qualification standards for miners are to be established by "consensus," essentially unanimity, which gives land-based producers as much influence as the United States. The possession of a veto provides them with an opportunity to extract potentially expensive concessions—new limits on production, for instance—to let the ISA function. Unfortunately, once the ISA asserted jurisdiction over seabed mining, potential producers would be hurt by a deadlock. . . .

TECHNOLOGY TRANSFER

Finally, there is technology transfer, one of the most odious redistributionist clauses of the original convention. The mandatory requirement has been discarded, replaced by a duty of sponsoring states to facilitate the acquisition of mining technology "if the Enterprise or developing States are unable to obtain" equipment commercially. Yet the Enterprise and developing states would find themselves unable to purchase machinery only if they were unwilling to pay the market price or were perceived as being unable to preserve trade secrets. The clause might be interpreted to mean that industrialized states, and private miners, whose "cooperation" is to be "ensured" by their respective governments, are then responsible for subsidizing the Enterprise's acquisition of technology. Presumably, the United States and its allies could block such a proposal in the Council, but again, it is hard to predict future legislative dynamics and potential logrolling in an obscure UN body. . . .

A Bad Treaty

All in all, the LOST remains captive to its collectivist and redistributionist origins. Nevertheless, the administration appears to sincerely believe that it has made a good bargain. "This is a very important development for the United States because it gives us an opportunity to, in a market-oriented context, gain the benefits of some of the resources of the Law of the Sea," explained Secretary Christopher to Congress. If only that were true. In fact, consumers worldwide are far more likely to gain the benefits of seabed minerals without the LOST.

Admittedly, the administration has made a bad treaty better. As Deputy Assistant Secretary Colson observed, the new accord "is quite an advance." But that does not mean the LOST is acceptable. Created at a time when statism held sway internationally, the LOST remains a bad agreement. It is simply beyond fixing. The Senate should reject the treaty, consigning it to the ash heap of history where it belongs.

"The deep ocean can be a destructive
environment for radioactive and
other hazardous wastes."

THE OCEAN FLOOR SHOULD NOT BE USED AS A NUCLEAR WASTE DISPOSAL SITE

Greenpeace

For many years beginning in 1946, the United States, the Soviet Union, and other nations routinely dumped nuclear waste into the ocean. A global ban of this practice was agreed to in 1993 by the London Convention, a consortium of seventy nations that regulates dumping at sea. However, some scientists have argued that the careful disposal of nuclear waste beneath the ocean floor might be the most environmentally benign solution to the problem of nuclear waste disposal and should be researched. In the following viewpoint, Greenpeace, an international environmental organization, argues that seabed disposal of nuclear waste must be rejected because it creates unacceptable risks to the ocean environment and because it is illegal under U.S. and international law.

As you read, consider the following questions:

1. What aspects of the ocean seabed environment make it unsuitable for nuclear waste disposal, in the opinion of Greenpeace?
2. According to Greenpeace, in what years did the United States begin and end its ocean dumping of nuclear waste?
3. What international treaties and agreements governing ocean dumping does Greenpeace mention?

Reprinted from Greenpeace, "Ten Reasons Why the Deep Ocean and Its Subseabed Are Inappropriate, Illegal Dumpsites for Radioactive/Other Hazardous Wastes," May 1997, by permission of Greenpeace. *References in the original have been omitted in this reprint.*

Ten reasons why the deep ocean and its subseabed are inappropriate, illegal dumpsites for radioactive/other hazardous wastes:

ENVIRONMENTAL REASONS

1. The oceans are a living, interconnected environment that can return radioactive and hazardous waste to humans via ocean food chains. Recent scientific discoveries confirm the existence of an immense diversity of animal life on and under the deep sea floor. The ocean also is a formidable environment, with pressures and temperatures reaching planetary extremes and legendary corrosive powers. The deep ocean can be a destructive environment for radioactive and other hazardous wastes. Whether these forces work on short or long time scales, the ultimate effect is to expose ever larger areas of ecosystem to materials initially deposited in a confined location. The objective of waste management should be to contain wastes so that they are isolated from the biosphere instead of diluting them, which is the case with ocean dumping, including seabed burial over longer time scales.

2. The oceans are continuous, with three broad interfaces: land-margins, the atmosphere, and the seabed. These interfaces are all critical areas of chemical and biological interactions so that the introduction of harmful materials at any interface can have far more profound effects than might be suggested by simply considering a localized disposal site. They are particularly vulnerable to change, by natural and human events, and relatively small actions can lead to relatively big consequences. Given the fluid nature of the medium and the interconnections and interfaces with other media, partial discontinuities, such as physical bottlenecks (e.g., straits), physico-chemical bottlenecks (e.g., vertical temperature layering), and horizontal differences in temperature, salinity and other physical and chemical qualities of sea-water (for living organisms), cannot be considered separately from each other.

3. In relation to the 1993 global ban on ocean dumping of radioactive wastes agreed under the London Convention (see para. 9, below), the parties carried out a detailed risk assessment by an Intergovernmental Panel of Experts on Radioactive Waste Disposal at Sea (IGPRAD). Completed in July 1993, that assessment provided a persuasive basis for the global, permanent ban. IGPRAD clearly identified a general regional and global trend to move away from ocean dumping, and agreed that the trend is justified since disposal at sea differs significantly from other disposal options. This is due in particular to the diffusibility of ra-

dionuclides in sea water which could result in transboundary transfers, and the inherent difficulty to monitor and—if necessary—retrieve wastes dumped in the ocean. . . .

DOMESTIC REASONS

4. The Ocean Dumping Ban Act of 1988 focused on ending U.S. industrial waste and sewage sludge dumping at sea, with clean dredged spoil material the only acceptable form of dumping now allowed. Under the Act, low-level radioactive waste dumping could only proceed if Congress enacts specific legislation authorizing such dumping. Even earlier, 1983 amendments to the Ocean Dumping Act of 1972 prohibited permits for low-level waste dumping unless designated findings are made by Environmental Protection Agency (EPA), and unless Congress adopts a joint resolution authorizing disposal within 90 legislative days thereafter.

5. For environmental reasons, the United States ended ocean dumping of radioactive wastes in 1970. Historically, the U.S. allowed dumping of radioactive waste at sites off our coasts from 1946–1970, with about 99% of that done prior to 1963. In 1970, however, the new Council on Environmental Quality CEQ) issued a report ('Ocean Dumping: A National Policy') which concluded that radioactive waste dumping at sea presented a serious and growing threat to the marine environment.

6. In 1976, special Congressional hearings were held to examine the question of radiological contamination of the oceans. At that time, senior policy analysts within the U.S. government were asked whether seabed burial of high-level radioactive waste (HLW) was dumping under the Ocean Dumping Act, and therefore prohibited. In a legal opinion issued by the lead agency responsible for administering the Act, the EPA specifically concluded that seabed burial was dumping, and that, as a result, seabed burial of HLW was prohibited since it was "black listed" from being dumped under the Ocean Dumping Act. . . . That legal opinion has served as the basis for the U.S. policy on seabed burial over the past 20-plus years.

7. The U.S. Department of Energy carried out a multi-year research and development program on seabed disposal during 1974–1986, primarily focused on HLW. That program was directed toward assessing the technical, engineering and environmental feasibility of the seabed disposal option. However, in 1986, the Department of Energy (DOE) terminated that program, stating that further assessments were unnecessary given the agency's preference for land-based disposal options.

INTERNATIONAL REASONS

8. Pursuant to Agenda 21, adopted by consensus at the 1992 Earth Summit, the 172 participating governments expressed support for the permanent ban on radioactive waste dumping at sea that was subsequently agreed under the London Convention in 1993 (see para. 9, below). The radioactive waste chapter of Agenda 21 included two pertinent action points, calling upon States to a) "expedite work to complete [the IGPRAD] studies on replacing the current voluntary moratorium on disposal of low-level radioactive wastes at sea by a ban, taking into account the precautionary approach . . ." and b) "not promote or allow the storage or disposal of high-level, intermediate-level and low-level radioactive wastes near the marine environment unless . . . scientific evidence . . . shows that such storage or disposal poses no unacceptable risk to people and the marine environment or does not interfere with other legitimate uses of the sea. making, in the process of consideration, appropriate use of the concept of the precautionary approach."

FUTURE ESTIMATES OF SPENT NUCLEAR FUEL

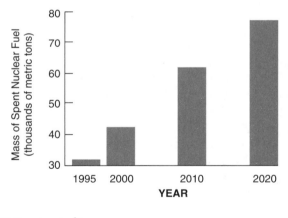

Source: U.S. Department of Energy

9. On November 12, 1993, the London Convention's treaty parties adopted global bans on ocean dumping of radioactive and industrial wastes, which came into force on February 20, 1994. Only two countries (Russian Federation, for the radioactive waste ban, and Australia, for the industrial waste ban) have not accepted the bans, having exercised their right under the Convention to opt out pursuant to amendment procedures—

though both are engaged in ongoing reviews that could lead to acceptance of the bans in the near future. Under the terms of the ban on radioactive waste dumping at sea, the continuing efficacy of that decision will be scientifically reviewed at 25-year intervals by the London Convention parties.

10. On November 7, 1996, the London Convention's parties adopted a revised treaty, building on the 1993 global bans. The new treaty—the 1996 Protocol to the Convention—was agreed at a 28 October–8 November 1996 Special Meeting of the Parties to the London Convention. Among other reforms, the parties agreed by consensus to codify the long-standing view of the vast majority of parties that seabed burial is dumping, defining "sea" to mean: "all marine waters other than the internal waters of States, as well as the seabed and the subsoil thereof . . . ," though it does not include subseabed repositories accessed only from land.

"If we investigate thoroughly, we just
might find that the best place to
deposit our high-level nuclear trash
is in the low-level abyssal clays
beneath the ocean."

THE OCEAN FLOOR SHOULD BE CONSIDERED FOR USE AS A NUCLEAR WASTE DISPOSAL SITE

Steven Nadis

In the following viewpoint, Steven Nadis argues that certain areas of the ocean floor may be potential disposal sites for radioactive waste produced by nuclear power plants and the manufacture and dismantling of nuclear weapons. He asserts that the careful burial of nuclear waste beneath the seabed should not be considered the equivalent of ocean dumping. Nadis encourages the United States and other countries to pursue a seabed burial strategy, and he criticizes U.S. government agencies for focusing exclusively on a land-based solution to nuclear waste disposal. Nadis is a science writer and former staff researcher for the Union of Concerned Scientists.

As you read, consider the following questions:

1. What exists beneath much of the ocean floor, according to Nadis?
2. In the author's view, why has the option of sub-seabed disposal been rejected by the U.S. Department of Energy?
3. What questions does the author list that should be addressed about sub-seabed nuclear waste disposal?

Excerpted from Steven Nadis, "The Sub-seabed Solution," *The Atlantic Monthly*, October 1996. Reprinted with permission.

In 1976 a giant coring device mounted to a ship plunged repeatedly into the bottom of the Pacific Ocean, three miles below the surface, bringing up 100-foot-long tubes of mud and clay with the consistency of peanut butter. The primeval muck told a tale of geologic serenity. Sediment records from the cores indicate that the region—roughly 600 miles north of Hawaii and spanning an area four times the size of Texas—has been tranquil for 65 million years, unperturbed by volcanic activity or by shifting of the earth's tectonic plates. Charles Hollister, a geologist and senior scientist at the Woods Hole Oceanographic Institution, saw even more when he gazed at the thick dark ooze. He saw what might prove to be the perfect place to sequester our high-level nuclear waste—the most potent and intensely radioactive by-products of military or civilian enterprise.

It's an intriguing vision, and one that in principle still holds great promise. Yet the concept of "sub-seabed disposal," first suggested by Hollister in 1973, has been undercut by a series of political blunders. . . .

THE IDEA

Hollister first hit upon the notion of sub-seabed burial in 1973, at a small social gathering in Washington, D.C. There he met William Bishop, a chemist at the Sandia National Laboratories, in New Mexico, who described the problems associated with a proposed nuclear-waste repository in Lyons, Kansas. "I immediately thought of the clays in the deep-sea floor, which I knew, from previous studies, clung tenaciously to the radioactive particles that had settled there as a result of atmospheric nuclear testing," Hollister recalls. He and Bishop stayed up all night discussing the idea, and a month later Hollister made a pitch to officials at Sandia, whose interest was piqued.

Next Hollister brought biologists, physicists, and oceanographers to Sandia to see if they could "destroy" the idea in what he calls the "biggest shootout since the OK Corral." He says, "If we could find out it was a stupid idea at the outset, it would save us a lot of time and money." But rather than shooting down the concept, many of the scientists told Hollister they'd like to work with him on it. A sub-seabed research program was initiated in 1974, with financial backing from Sandia; within a few years it had grown into an international effort involving ten countries and 200 scientists, under the auspices of the Paris-based Organization for Economic Cooperation and Development. This collaboration led to the core-sampling expedition that demonstrated the stability of the region underlying the

North Pacific floor. Hollister points out that the Pacific site he and his colleagues explored twenty years ago is not unusual, geologically speaking. "About a quarter of this planet is covered with geology that is appropriate for this solution," he says.

Experiments conducted by this international team of scientists from 1974 to 1986 support Hollister's opinion that the sticky mud and clays that blanket the mid-ocean basins may provide the best burial grounds yet proposed for nuclear waste. These tests suggest that if waste canisters were deposited just ten meters below the ocean floor, any toxic substances that leaked out would be bound up by the clays for millions of years. Deeper interment, at 100 meters or more, could easily be managed, providing an even greater margin of safety. "The stuff sticks to the mud and sits there like heavy lead," Hollister maintains. "Nothing's going to bring it into the biosphere, unless we figure out how to reverse gravity."

GOVERNMENT ROADBLOCKS

If he's right, and the proposed technique could end the worldwide radioactive-waste problem that has been building up for the past fifty years, why has almost nobody in this country heard about it? The answer to this question—along with the roots of many of the problems plaguing current U.S. nuclear-waste-disposal efforts—can be traced to a 1986 decision by the Department of Energy (DOE) which cut off research funds for sub-seabed and other disposal alternatives, so that the agency could focus exclusively on developing a land-based geologic repository for high-level wastes; a year later it settled on Yucca Mountain, Nevada. The timing was unfortunate: ongoing sub-seabed experiments were canceled in spite of encouraging results and after much experimental apparatus had already been built.

The federal government had a change of heart in 1987, when Congress passed amendments to the Nuclear Waste Policy Act which, among other things, established the Office of Subseabed Disposal Research within the DOE. The director of this office, Walter L. Warnick, was asked to create a consortium of university investigators and devise a long-range research plan. But a couple of months after Warnick had enthusiastically begun, the congressional committee that controlled appropriations strongly discouraged the Energy Department from spending any money on the program. With access to sub-seabed research funds blocked, Warnick shifted his attention to acid rain and global-warming issues. The Office of Subseabed Disposal existed in name only until 1996, when it was abolished altogether. . . .

The sub-seabed approach has been the subject of peer-reviewed research, and the program generated dozens of articles in prominent international scientific journals. Henry Kendall—a Nobel laureate in physics, a professor at the Massachusetts Institute of Technology, and the chairman of the Union of Concerned Scientists—calls sub-seabed disposal a "sweet solution" and a "winner," labeling it the best of the alternatives from a technical standpoint. A National Academy of Sciences panel called for further study of the sub-seabed approach, and a 1995 report by Robert Klett, a systems analyst at Sandia, concluded that "[all] analyses to date indicate that sub-seabed disposal would be a safe and economical method of [high-level waste] disposal and that predictions could be made with a high degree of confidence." In light of these endorsements, why isn't the idea being pursued, if only through research? Why won't this country make the modest investment— about ten years and $250 million, according to Hollister—required to find out if it would really work?

PROBLEMS AT YUCCA MOUNTAIN

The reasons are varied, though they are woven together in a familiar pattern. The Department of Energy killed the program partly for political reasons and partly because the sub-seabed researchers never really fit in with mainstream DOE culture. "It was a clear case of 'not invented here,'" Hollister says. Many environmentalists—acting as narrow-mindedly as their traditional opponents in government and the nuclear industry—dismissed the idea before learning the details, assuming that the approach involved little more than wholesale ocean dumping. The nuclear utilities lobbied against it for pecuniary reasons: the waste-disposal effort is largely subsidized by a tax on nuclear-generated electricity that the utilities have been paying (they pass the cost on to consumers) since 1982, and they have seen little tangible return on their $12 billion investment. Industry officials—concerned that the DOE would be unable to meet its obligation to start accepting nuclear waste by 1998—surmised that the sooner the Yucca Mountain facility opened, the sooner they could divest themselves of their spent nuclear fuel and the waste issue in general. "Their position was extremely superficial," says John Kelly, who heads JK Research Associates, a consulting firm specializing in nuclear- and hazardous-waste disposal issues. "They decided the only way to succeed in building a repository in Nevada was to cut off all alternatives." This position was shared by Louisiana Senator J. Bennett Johnston, then the chairman of the Energy and Natural Resources Committee and a leading opponent of sub-seabed disposal.

The shortcomings of the resultant program are now widely apparent. After spending almost $2 billion on technical studies and preliminary excavation at Yucca Mountain, the DOE still hasn't demonstrated the geologic suitability of the site. The mountain lies near active seismic faults and a volcano that erupted less than 10,000 years ago. There is concern that the water table beneath the proposed burial grounds could rise and seep into the repository, contaminating groundwater and allowing radioactivity to escape. Two scientists, holding a decidedly minority view, have even suggested that the buried wastes might "go critical" and explode because of the large amounts of fissionable material packed into a relatively small space. Meanwhile, political opposition is growing: Nevada's governor and senators, along with local environmental groups, have declared war on the venture. Even if the project can withstand these challenges and move forward, the facility will be seriously undersized the day it opens its doors (2015 is considered the earliest possible date), able to accommodate just a fraction of the high-level waste that will have accumulated by then in the United States.

SUBSEABED DISPOSAL

Subseabed disposal . . . would utilize some of the world's most stable and predictable terrain, with radioactive waste or nuclear materials from warheads "surgically" implanted in the middle of oceanic tectonic plates. Selecting sites for disposal that are far from plate boundaries would minimize chances of disruption by volcanoes, earthquakes, crustal shifts and other seismic activity. Many studies by marine scientists have identified broad zones in the Atlantic and Pacific that have remained geologically inert for tens of millions of years. What is more, the clay-rich muds that would entomb the radioactive materials have intrinsically favorable characteristics: low permeability to water, a high adsorption capacity for these dangerous elements and a natural plasticity that enables the ooze to seal up any cracks or rifts that might develop around a waste container.

Charles D. Hollister and Steven Nadis, *Scientific American*, January 1998.

The government's unwillingness to prepare a good fallback position in the face of mounting difficulties seems like sheer folly. Although the DOE is not supporting any work on alternative disposal concepts at present, Hollister has not given up. While the ambitious research program he helped to fashion is on hold, he continues to explore the sub-seabed concept in indirect ways. In 1993, for example, he spent six weeks in the Norwegian Sea studying a Soviet nuclear sub that had sunk years before in the middle of an active fishing ground. "The scientific evidence to date points to zero impact if the nuclear material sits beneath the bottom of the sea or even on the bottom," he says. Other analyses of radioactive spills in the marine environment have reached a similar conclusion: high-level radioactive materials tend to stay put if they are placed in or on a clay-rich sea floor, Hollister claims. Vertical migration rates are so slow that it is "virtually impossible" for measurable concentrations of radioactivity to reach the surface from deep water. "Many people don't like this conclusion," he adds, "but I've never seen any data in the oceanographic literature that refute it."

He concedes that the sub-seabed-disposal concept requires additional scrutiny. Several questions, all fairly straightforward, still need to be addressed. For instance, is it better to put the waste in torpedo-shaped canisters that will penetrate the sea floor on their own after being dropped from the surface, or should it be implanted by means of drilling? How deep should the canisters be buried? How will the heat generated by radioactive waste affect the muds lying beneath the ocean? And how strongly will negatively charged clay particles latch on to positively charged ions of uranium, plutonium, cesium, and strontium? "We know exactly what to do to answer these questions," Hollister says, citing field experiments that have already been designed to determine, for instance, how securely the holes close up around the waste canisters and whether radioactive material moves through ocean-floor clays at the same rates measured in the lab.

Arrangements would have to be made, of course, to ensure the safe transport of waste to the seabed. International laws governing the use of the seas would also have to be addressed, although this might be done while research is under way.

THE LONDON DUMPING CONVENTION

A forum already exists that can resolve issues related to sub-seabed disposal: the annual meetings of the contracting parties to the London Dumping Convention, the only international treaty that directly addresses the subject. Should sub-seabed nuclear-waste disposal ever be implemented, the program could be administered and regulated worldwide by the International Maritime Organization, the agency that oversees the functions of the LDC. This prospect may be especially attractive to countries like the Netherlands and Japan, which have little room and no favorable geology for land-based disposal methods.

Yet the treaty may pose the biggest hurdles for this waste-disposal option. Although the dumping of any radioactive waste at sea has been prohibited by international law since 1994, the status of sub-seabed disposal has been ambiguous. This may change very soon: a resolution to be taken up at an LDC meeting at the end of October 1996 would extend the definition of "dumping" to include "any deliberate disposal or storage of wastes or other matter in the sea-bed and the subsoil thereof." If the measure passes (and indications are that it will), sub-seabed disposal will be prohibited, and the decision may not be appealed for twenty-five years. [Editor's note: Seabed disposal of nuclear waste was declared to be illegal ocean dumping at the October 28–November 8, 1996 meeting of the London Dumping Convention.]

The resolution makes no sense to Edward Miles, an expert on international marine policy at the University of Washington. "On objective grounds, there is no way one can argue that sub-seabed burial is dumping," he says, pointing out that the United Nations' International Atomic Energy Agency considers it a "variation of deep geologic disposal on land." The United States, which lobbied for the sub-seabed approach at the 1984 LDC meeting, has since "flip-flopped" on the issue, he says, and now supports the resolution. "It's not a position based on any scientific or legal matters. It's just a political decision attributable to the fact that the environmental community has access to Al Gore and, through him, to President Clinton."

ARGUMENTS AGAINST SEABED DISPOSAL

Clifton Curtis, a political adviser to Greenpeace International, has fought against sub-seabed disposal since 1978, at LDC meetings and elsewhere, in his campaign to "protect the ocean from potentially harmful activities." He favors land-based disposal options, arguing that "the people who produce nuclear waste should deal with it in their own territory—that would force everyone to pay more attention to what they're producing." Terrestrial methods are also superior on the grounds of "monitorability" and "retrievability," he says. "If there's ever the need to retrieve the wastes because of a problem, it's much easier to do so on land."

Hollister disputes this contention, asserting that the technology exists for both monitoring and recovery at sea. What's more, he says, sub-seabed retrieval would probably be easier and cheaper than digging vast tunnels into the earth. Yet Curtis remains unconvinced and would like to see the impending ban on sub-seabed disposal apply to research as well—a viewpoint, he claims, that is shared by all major environmental organizations. "There is a broad consensus that ocean dumping of radioactive wastes, including sub-seabed burial, should be prohibited," he says. "In light of that, it makes more sense to focus our research on terrestrial options."

The London treaty takes no formal position regarding experiments that would involve putting small amounts of radioactive material in the ocean floor. But even if such studies are not explicitly forbidden, Miles argues, the U.S. government is highly unlikely to fund such research in the face of an international (and national) disposal ban. "A decision to classify sub-seabed disposal as dumping would effectively kill the idea by cutting off any motivation to continue the research," he says. "And without any additional research there will be nothing to reconsider twenty-five years from now.". . .

A PUZZLEMENT

Hollister is puzzled by the flurry of attacks on a field that has been unfunded and dormant for a decade. "I have no problem with a ban on sub-seabed disposal," he says. "I think it should be banned until we do more experiments. What troubles me is people who are trying to ban *research* on the subject." Rather than discouraging inquiry to meet short-term political objectives, he says, we should be actively exploring all reasonable disposal options. He suspects that the opposition to sub-seabed disposal is part of a broader opposition to nuclear energy in general. "Some people don't want to hear about anything nuclear, even solutions," he says. . . .

If we investigate thoroughly, we just might find that the best place to deposit our high-level nuclear trash is in the low-level abyssal clays beneath the ocean. It's possible, of course, that we'll conclude after further study that sub-seabed disposal is not the answer, but we ought to spend the time and money to find out.

PERIODICAL BIBLIOGRAPHY

The following articles have been selected to supplement the diverse views presented in this chapter. Addresses are provided for periodicals not indexed in the *Readers' Guide to Periodical Literature*, the *Alternative Press Index*, the *Social Sciences Index*, or the *Index to Legal Periodicals and Books*.

David Colson	"U.S. Accession to the UN Convention on the Law of the Sea," *U.S. Department of State Dispatch*, February 13, 1995.
Charles D. Hollister and Steven Nadis	"Burial of Radioactive Waste Under the Seabed," *Scientific American*, January 1998.
Peter Leitner	"A Bad Treaty Returns: The Case of the Law of the Sea Treaty," *World Affairs*, January 1, 1998.
Richard Monastersky	"A Nuclear Waste Experiment at Sea?" *Science News*, January 4, 1997.
Jim Motavalli	"Nuclear Waste: A Watery Grave?" *E Magazine*, January/February 1998.
Seydou Amadou Oumarou	"Sea Tribunal Makes a Slow Start," *UNESCO Courier*, July/August 1998.
William Safire	"LOST at Sea," *New York Times*, March 31, 1994.
Michael Satchell	"Lethal Garbage: Nuclear Waste," *U.S. News & World Report*, November 7, 1994.
David Schneider	"Not in My Backyard," *Scientific American*, March 1997.
U.N. Chronicle	"The UN Convention on the Law of the Sea: A Chronology," March 1995.
U.S. Department of State Dispatch	"Fact Sheet: U.S. Oceans Policy and the Law of the Sea Convention," March 11, 1996.
Susan Wells and Gordon Shepherd	"Threats and Conflicts," *UNESCO Courier*, July/August 1998.
Tim Zimmermann	"If World War III Comes, Blame Fish," *U.S. News & World Report*, October 21, 1996.

WHAT POLICIES WOULD BEST PROTECT THE WORLD'S FISHERIES?

CHAPTER PREFACE

Humans have fished the oceans for thousands of years using hooks, spears, nets, and traps. However, modern commercial fishing has greatly changed the way fish are caught. Sonar detectors can locate fish deep under water. Satellite navigation systems can help fishing boats pinpoint the spots where schools of fish gather. Commercial fishers use hooked "longlines" that can extend for miles and extremely large and light nylon nets capable of taking 130 tons of fish in a single haul. Because of technological advances and the growth of the world's fishing fleet, the annual total fish catch grew steadily from 18.5 million tons in 1952 to 89 million tons in 1989.

Since 1989, however, the total wild fish catch has leveled off at about 90 million tons and has declined in some years even as the number of fishing ships has continued to grow. Some famous fisheries, such as the Grand Banks cod fishery off Newfoundland, Canada, have collapsed entirely, forcing fishers to shut down operations or to catch and sell previously undesirable species. Whether such local problems are emblematic of a worldwide crisis is somewhat in dispute. Commercial fishers, marine biologists, and government regulators often disagree over the health of fish populations. Disputes also exist over the causes of fish declines, which are thought to include changes in ocean currents, pollution, and destruction of coastal habitats. Many observers contend that a main reason is simply overfishing—taking too many fish for their population to replenish naturally.

Many people have called for regulations limiting technology applied to fishing. For instance, the environmental organization Greenpeace has argued for restrictions on factory trawlers—large mechanized vessels capable of catching five hundred tons of fish a day. In a 1998 report the organization asserted: "There are simply too many large-scale, high-tech fishing boats roaming the world's oceans on an unsustainable course of plunder for profit wherever fish stocks can be found." Commercial fishers argue that such restrictions would threaten their livelihoods and would not prevent overfishing. The At-Sea Processors Association (ASPA), a fishing industry group, has accused Greenpeace of "manipulating information and public emotion to create a fundraising issue." The viewpoints in this chapter examine this and other debates over preserving the world's ocean fisheries.

"Factory trawlers . . . catch and waste
millions of pounds of sea life every
year."

THE ENVIRONMENTAL CASE FOR
BANNING FACTORY TRAWLERS

Greenpeace

The following viewpoint advocating the elimination of factory
trawlers is taken from a position paper by Greenpeace, an international environmental organization. Greenpeace argues that factory trawlers—large ships in which fish are both caught and processed—pose a major threat to the sustainability of the world's
fisheries because of their ability to catch fish in large numbers. It
asserts that the large pollock fish population in the Bering Sea off
of Alaska, for instance, is endangered by a factory trawler fleet
whose capacity exceeds limits set to maintain the fishery's population. The organization urges Congress to enact legislation banning the use of factory trawlers in U.S. territorial waters.

As you read, consider the following questions:

1. How many fish can factory trawlers catch in a single haul,
   according to Greenpeace?
2. What provisions of S. 1221, the bill introduced by Senator
   Ted Stevens of Alaska in September 1997, does Greenpeace
   describe?
3. What other forms of marine wildlife besides fish does
   Greenpeace argue are being hurt by overfishing?

Abridged from "Banning Factory Trawlers: The Environmental Case for S. 1221," a
position paper of Greenpeace International. Reprinted by permission of Greenpeace.

Fishing is an ancient human tradition, and for most people, conjures up images of fishermen braving the elements to catch a few fish for market. But in the last 50 years, fishing has been transformed into a hi-tech, global industry that has the power to radically alter marine ecosystems. All around the world, overfishing and destructive fishing practices are destroying fish stocks, damaging marine ecosystems and threatening the livelihoods of tens of millions of people.

One of the primary causes driving the fisheries crisis is the fact that the capacity of the world's fishing fleet to catch fish greatly exceeds the amount of fish that can be caught on a sustainable basis. In other words, there are simply too many boats, especially large-scale, industrial vessels, like factory trawlers, chasing fewer and fewer fish. As a result, many fish stocks are dangerously depleted, and entire ecosystems are being turned upside-down. The threats posed by excess fishing capacity to the world's fish stocks have been acknowledged in major international agreements, including the United Nations (UN) Agreement on the Conservation and Management of Highly Migratory Fish Stocks and Straddling Fish Stocks and the UN Food and Agriculture Organization (FAO) Code for Responsible Fishing.

In the United States, the problems created by excess fishing capacity are nowhere better exemplified than in the country's largest fishery for Alaska pollock, where the glut of factory trawler capacity exceeds the total allowable catch limit by at least two to three times. The factory trawler fleet, which has ranged between 45 and 65 vessels in recent years, catches nearly a full one-fifth of the total U.S. catch each year. Some 36,000 smaller fishing vessels account for the rest of the catch.

GLOBAL SEA MONSTERS

Factory trawlers are large, industrial fishing vessels that catch fish in huge nets, some as long as four football fields. The largest nets can capture 400 tons of fish in a single haul. A factory trawler can stay out at sea for months at a time without returning to port—fishing, processing and storing fish on board, around-the-clock, seven days a week. Factory trawlers are known for producing vast quantities of bycatch, the unwanted fish and other sea life caught and thrown overboard, dead or dying. In 1994, U.S. factory trawlers in the North Pacific discarded 260,000 metric tons of bycatch, exceeding by 42% the total groundfish landings off New England that year. Mobility permits factory trawlers to easily abandon depleted waters for fertile fishing grounds elsewhere in the world—leaving local fishers to face the consequences of a wasted ecosystem.

Recognizing the devastating environmental potential of factory trawlers, Congress enacted measures in 1997 to prohibit the entry of large-scale fishing vessels into two important East Coast fisheries for Atlantic herring and mackerel. The catalyst for this legislation was a 369-ft. factory trawler which had set its sights on these two fisheries. This legislation was passed with significant support from both environmental and fishing industry groups.

THE AMERICAN FISHERIES ACT

In September 1997, Senator Ted Stevens (R-AK) introduced legislation that would begin to tackle the issue of excess capacity in the North Pacific, by starting to phase out the current U.S. factory trawler fleet. Not only would enactment of S. 1221 be a logical extension of 1997's Congressional action, but moreover, a clear demonstration of U.S. leadership on the fundamental problem of excess fishing capacity.

If enacted as introduced, S. 1221 would have:

- made permanent the current three-year moratorium on subsidies to build new large-scale fishing vessels, including factory trawlers;
- prohibited the entry of additional U.S. or foreign factory trawlers into any U.S. fishery;
- required the sale of any factory trawler that does not meet the 75 percent U.S. ownership requirements contained in the bill;
- forbidden the replacement of any remaining factory trawlers at the end of their useful life. [Editor's note: S. 1221 was not passed by Congress as of August 1998.]

U.S. FACTORY TRAWLERS

Modern factory trawlers are very expensive, some costing as much as $40 million. The more expensive a ship is, the more fish it will need to catch to remain profitable. Debts create financial pressures which are incompatible with sustainable fishing practices. Excess fishing capacity and debt-driven economics encourage wasteful, dangerous, and destructive fishing practices such as fishing heavily on spawning stocks in pursuit of the lucrative roe and "pulse fishing," where a stock or local school of fish is set upon and fished out.

In the North Pacific, a dramatic surge of factory trawler capacity, and influx of capital, has had profound effects on the conduct of the fishery and the health of the marine environment. Management efforts to slow down the harvest, lower annual catch quotas, spread out fishing effort, and reduce bycatch are all undermined by this surplus capacity. Unfortunately, the result has been that the North Pacific Fishery Management Council spends much of its time and resources engaged in costly allocation deliberations in an attempt to divide the Alaska pollock quota and prevent the factory trawl fleet from preempting shore-based catcher boats.

1. *Overcapitalization and Excess Capacity*
- The domestic factory trawl fleet did not exist prior to 1983. It grew rapidly from 12 vessels in 1985 to 45 by the end of 1988, while catching capacity grew from 250,000 metric tons to 800,000 tons.
- Additions to the fleet in the 1988–1990 period added 15 large pollock surimi/fillet factory trawlers with a production potential estimated to be "several times the size of the vessels which entered the fleet between 1985–1987." These vessels were rebuilt in foreign countries after the passage of the Anti-Reflagging Act (1988), which Congress enacted to halt the entry of more factory trawlers into the North Pacific. However, all of these vessels ultimately received exemptions from the Coast Guard under the Act's grandfather clause.
- By 1991, 50 factory trawlers in the pollock fishery, comprising only 2.5% of all groundfish vessels that year, caught over 1 million metric tons of Alaska pollock—three-quarters of the Bering Sea pollock quota.
- By 1992, there were some 65 factory trawlers and an estimated $1.6 billion investment in the fleet.
- In 1994, 24 of largest class of surimi factory trawlers comprised only 1.5% of the total number of groundfish vessels fishing in the Bering Sea/Aleutian Islands yet they accounted for 30% of the entire groundfish catch.

2. *Shortened Seasons*
The dramatic curtailment of the Bering Sea pollock season reflects the impact of adding so much factory trawl capacity in the 1990s:
- In 1989 the pollock fishery was still open year-round. In 1991, the season was reduced to only 148 days. By 1994, the factory trawl season declined only 70 days long. In 1997, the factory trawlers fished for 55 days.
- When the pollock season was reduced to three months or less, many boats could not make payments on their loans. Most continue to fish every year, however, thanks to buy-

112

outs by wealthier competitors, particularly by the Norwegian subsidiary American Seafoods.

- Shorter seasons, combined with extreme competitive and financial pressures, foster destructive and unsustainable fishing practices. As the race to catch fish intensifies, vessel owners are forced to upgrade their boats to become more efficient and competitive, investing millions of dollars in the process. In the end, the need to ensure a return on those investments often wins out over the need to protect the fish for the future.

3. Overfishing

The economics of excess capacity and overcapitalization drive factory trawlers to fish as hard as possible. Factory trawlers require high volumes of fish to remain profitable. Unfortunately, in response, the North Pacific Fishery Management Council appears compelled to keep annual quotas high in an attempt to accommodate this fleet:

- Total pollock catches in the Bering Sea/Aleutian Islands from 1990–96 were 8,786,189 metric tons, an average of 1.255 million metric tons annually not including discards. These figures are higher than the historical average, higher than any catch total in the 1980s and higher than any recorded catch in the eastern Bering Sea since 1975.

Despite National Marine Fisheries Service's claims to the contrary, there are myriad signs that several key commercial stocks are being overfished in a single-species context:

- Aleutian Island pollock (age 3+ biomass) has declined steadily since the early 1980s and appears to be at about 20% of its earlier abundance.
- Eastern Bering Sea pollock has declined by more than half since the mid-1980s and by 38% from 1994 to 1997. . . .
- Until the 1990s, the Aleutian Atka mackerel catch never exceeded 38,000 metric tons. Beginning in 1992, the North Pacific Council raised the annual catch limits to 43,000 tons, increasing to 80,000 tons by 1995 and 106,157 tons in 1996. In 1997, the estimated stock size was down 50% compared to the 1991–1994 time period, suggesting that these record-high catch levels are not sustainable, and are driving down the stocks.

ECOLOGICAL EFFECTS OF FACTORY TRAWLERS

Examining these fisheries from the perspective of the ecosystem, it is clear that overfishing is occurring in an ecosystem context:

- Pollock is the most abundant groundfish species in the

THE SCALE OF MODERN FISHING

What has amplified the destructive power of modern fishing more than anything else is its gargantuan scale. Trawling for pollock in the Bering Sea and the Gulf of Alaska, for example, are computerized ships as large as football fields. Their nets—wide enough to swallow a dozen Boeing 747s—can gather up 130 tons of fish in a single sweep. Along with pollock and other groundfish, these nets indiscriminately draw in the creatures that swim or crawl alongside, including halibut, Pacific herring, Pacific salmon and king crab.

J. Madeleine Nash, *Time*, August 11, 1997.

Bering Sea, comprising 50–70% of the entire groundfish biomass, and a primary prey species for many marine mammals and seabirds in the region. The declines of major pollock predators, such as the Steller sea lion, in heavily fished areas of the Bering Sea, Aleutian Islands and Gulf of Alaska suggest that large changes in the ecosystem have occurred since the advent of factory trawlers and industrial-scale fishing.

• The cumulative effects of modern factory fisheries off Alaska may be a limiting factor in the recovery of declining Bering Sea wildlife populations. According to the National Research Council's 1996 Report on the Bering Sea Ecosystem, "It seems extremely unlikely that the productivity of the Bering Sea ecosystem can sustain current rates of human exploitation as well as the large populations of all marine mammals and birds that existed before human exploitation—especially modern exploitation—began."

• The northern fur seal population was listed as depleted under the Marine Mammal Protection Act in 1989. Large declines in Steller sea lions in western Alaska led to their listing under the Endangered Species Act as threatened in 1990 and endangered in 1997. Similarly, large declines have been documented in the harbor seal populations of western Alaska. Substantial declines of pollock-eating seabirds, specifically murres and kittiwakes, in the Bering Sea have been recorded beginning in the mid-1970s and coincident with the growth of the pollock fishery in the region.

• The National Research Council's 1996 report identified fishery effects on prey availability as the only factor with a high likelihood of playing a major role in the Steller sea lion's on-going decline.

4. Increased Roe Fishing

By 1994, factory trawlers in the Bering Sea were deriving more than one-fourth of their annual revenues from lucrative pollock roe. For many, financial success has hinged on the outcome of the roe fishery. Although roe-stripping—retaining only the roe, and discarding the fish carcass—has been officially prohibited in the pollock fishery, the volume of roe fishery is unprecedented. The declining size of the Bering Sea spawning stock over the course of the 1990s makes the roe fishery even riskier, since the reproductive population is most vulnerable to trawl gear at spawning time. . . .

5. High Volumes of Bycatch and Discards

Huge, indiscriminate nets, and physical limitations of onboard processing equipment and storage space, force factory trawlers to catch and waste millions of pounds of sea life every year. In 1998, new rules requiring that all pollock and cod caught be retained will prevent this form of waste, but will not reduce the volume of lethal bycatch: it will simply mean that these fish are ground up into fish meal:

- Overall, factory trawlers catch more unwanted or unusable fish, and throw more of it away, than any other vessel class in the groundfish fisheries off Alaska, accounting for more than three-quarters of total Bering Sea/Aleutian Islands discards in 1994. The entire factory trawl fleet's discards off Alaska in 1994—572 million pounds in all—was more than 3 times the reported discards (170 million pounds) of the more than 2,000 other boats that fished for groundfish that year.
- Factory trawlers off Alaska discard roughly two to three times more bycatch than the Alaska-based catcher boat fleet.
- From 1990–94, the total tonnage of discards in the Bering Sea/Aleutian Islands groundfish fisheries ranged between 197,660 and 314,585 metric tons.
- Even in the Bering Sea midwater pollock fishery, where the bycatch/discard rate is low, the total volume of discarded pollock and other species has been the highest of any Bering Sea groundfish fishery, averaging more than 93,000 metric tons (205 million pounds) per year from 1990–94.
- Total discards for yellowfin sole and Pacific cod averaged 60,000 metric tons per year from 1990–94, second only to the pollock fishery.
- Bycatch and discards rates are much higher for the factory trawl fleet fishing for yellowfin sole, rock sole and other

bottom-dwelling fish, including Atka mackerel, Pacific cod, and flathead sole. In 1997, the Bering Sea/Aleutian Island discard rate for yellowfin, flathead and rock sole—all carried out primarily by factory trawlers—ranged from 35–55%.

- The rock sole fishery has had the highest rate of total discards, ranging from 60–70% from 1990–94.
- Trawling on hard-bottom habitat in the Aleutian Islands by factory trawlers fishing for Atka mackerel may have destroyed slow-growing, cold-water corals that were once a major component of bycatch in the fishery. After 20 years, the corals are infrequently found in areas where trawling has been most intense.

> "The hypothesis . . . that large fishing
> vessels are incompatible with sound
> fisheries conservation . . . is not
> supported by the facts."

FACTORY TRAWLERS SHOULD NOT BE BANNED

Paul MacGregor

Paul MacGregor is executive director of the At-Sea Processors Association (APA), an industry group representing American factory trawlers that fish in the Pacific Ocean. In the following viewpoint, taken from his testimony before Congress, MacGregor speaks in opposition to proposed legislation that would ban ships larger than 165 feet from operating in U.S. waters because of their alleged negative impact on the environment. According to MacGregor, the factory trawlers in the North Pacific have an excellent environmental record. Furthermore, he asserts that there is no connection between a ship's size and its capability to fish with adequate conservation practices—and that smaller ships are just as capable as larger ones of exploiting a fishery to the point of collapse.

As you read, consider the following questions:

1. According to MacGregor, what are the keys to effective fisheries management?
2. What reasons does the author give to account for the size of factory trawlers?
3. What safety issues are raised by banning large factory trawlers, according to MacGregor?

Excerpted from Paul MacGregor's testimony before the Subcommittee on Oceans and Fisheries, Committee on Commerce, Science, and Transportation, U.S. Senate, on S. 1221, The American Fisheries Act, March 26, 1998.

At-Sea Processors Association (APA) represents companies that operate twenty-four U.S.-flag at-sea fish processing vessels. APA's catcher/processors, or factory trawlers, are principally engaged in the Bering Sea pollock fishery and the West Coast Pacific whiting fishery. Over 90 percent of the fleet's revenues are derived from those two fisheries. Both of these fisheries are primarily conducted with mid-water trawl gear and are widely recognized as two of the cleanest fisheries in the world. To a much lesser degree, some APA member vessels also participate in one or more other Bering Sea groundfish fisheries.

The fleet contributes substantially to the Pacific Northwest and Alaska fisheries and maritime economies. The at-sea pollock processing fleet employs nearly 4,000 workers. Two-thirds of the workers reside in Washington State. Alaska, Oregon and California residents are also strongly represented in the workforce. . . .

CONSERVATION ISSUES

The hypothesis advanced in S. 1221 [proposed legislation banning factory trawlers]—that large fishing vessels are incompatible with sound fisheries conservation and, therefore, should be phased out—is not supported by the facts. Most U.S. fishing vessels longer than 165 feet fish in the North Pacific. A September 1997 National Marine Fisheries Service (NMFS) report on the status of U.S. fish stocks reports found that of the 63 fish species identified in the North Pacific, none are overfished. In fact, the biomass of Bering Sea groundfish alone has increased from 11 million metric tons to 17 million metric tons since 1980. Contrast that situation with the overfished East Coast and West Coast groundfish fisheries that are the province of small vessel fleets, and it is clear that improvements in fisheries management will not result from imposing a federal vessel size limit. . . .

There is ample evidence in U.S. and world fisheries that vessels of any size, when not properly regulated, can over-harvest fishery resources. Effective and enforceable fisheries management regimes are essential if conservation goals are to be met. The North Pacific groundfish fishery is an example of perhaps the best managed fishery in the world. It could even be argued that having fewer, larger vessels in that fishery provides conservation benefits because fishery managers are better able to monitor and enforce fisheries regulations.

The keys to effective fisheries management in the North Pacific are:

• Federal scientists determine the population abundance for

each species and establish safe harvest guidelines on an annual basis.

- Fishery managers establish a catch limit for each species of groundfish at or below the sustainable level recommended by scientists. When the quota is taken, the fishery shuts down.
- All catch, whether retained or discarded, counts against the allowable catch level.
- All pollock catcher/processors carry onboard at least one federal fishery observer to monitor and record catch composition and amounts. Smaller vessels have lower levels of observer coverage.
- State of the art electronic reporting systems allow fishery managers to monitor catch levels and manage quotas on a real time basis.

BENEFITS OF LARGE FISHING VESSELS

Despite a solid record of sound resource management, overcapitalization remains a root problem in the North Pacific groundfish fishery. Both the onshore and offshore fishing and fish processing sectors have invested too much capital in an effort to win the so-called race for fish. Both sectors are overcapitalized. However, since catch limits are set at conservative levels and are strictly enforced, overcapitalization is an economic concern in the pollock fishery, not a conservation concern. The findings of S. 1221 allege, without supportable evidence, that overcapitalization in the at-sea sector will lead to pressure on fishery managers to set harvests above sustainable levels. If that was the case, overcapitalization in the shoreside sector and in other fisheries would lead to similar results. However, neither APA, nor any of its member companies, has ever advocated catch levels above the allowable biological catch (ABC) level recommended by the scientists. Instead, we have consistently urged federal fishery managers to end the race for fish and to rationalize the North Pacific fishery. Regrettably, little progress is being made in that regard.

It should also be recognized that vessel length is not necessarily indicative of harvesting capacity. Although U.S.-flag pollock factory trawlers range in length from 220 to 385 feet, catcher vessels half the length of factory trawlers often have comparable fishing power. For example, many factory trawlers and trawl catcher vessels employ similar sized nets. This is borne out by NMFS data showing that tow sizes, that is the amount of fish harvested from each set of a net, are roughly the same for catchers and catcher/processors in the Bering Sea.

HARMING AN IMPORTANT INDUSTRY

There is no basis for Greenpeace's demand that $1.5 billion worth of capital investment by fishermen be scrapped by the year 2001. The factory trawler fishing fleet was built because American fishermen responded to Congressional policies that foreign fishermen and processors operating in U.S. waters be phased out and replaced by an American fishing fleet. Some American fishermen responded to this emphasis in federal policy by building factory trawlers, that is, placing fish processing equipment aboard their vessels and adding value to the products that they produce. Harvesting species of fish that, heretofore, only foreign fishermen caught, American fishermen not only became processors of their own catch, but marketers as well. Substantial trade barriers had to be overcome before they gained entry to certain overseas markets. Now, exports of North Pacific groundfish products are valued at $500 million annually.

The U.S. factory trawler fleet employs 10,000 men and women. All of these jobs were created in the 1980s when this new fleet of vessels was built. With its vicious and unsubstantiated attacks, Greenpeace does great harm to the workers—and to the families of the workers—whose livelihoods are tied to the North Pacific fisheries.

At-Sea Processors Association, "The Greenpeace Deception: Executive Summary," 1996.

The purpose and function of a 300-foot-long pollock factory trawler is to accommodate onboard processing equipment, crew quarters for 75 to 125 men and women, galley facilities, a cold storage hold, and, of course, fishing equipment. Fishermen use at-sea processing technology as a means of adding value to their catch. After all, unprocessed pollock is worth only 8 cents a pound; whereas, pollock fillets processed at-sea currently sell for about $1.30/pound. A phaseout of large fishing vessels would effectively preclude an option for fishermen to continue to process, and add value to, their own catch in this fishery. They would be relegated instead to delivering their catch to one of a handful of processors and being offered pennies a pound for their fish. . . .

BYCATCH AND WASTE ISSUES

The findings and rationale of S. 1221 assert that fishing vessels greater than 165 feet in registered length are "less likely than smaller, less powerful vessels to avoid bycatch and waste" in the fishery. Alaska Department of Fish and Game (ADF&G) statistics are cited to show that "fifty-five factory trawlers in the Bering Sea threw overboard 483 million pounds of groundfish, wasted and unused in 1995." No effort is made, however, to correlate the bycatch and discard rates cited with the vessels that are most adversely affected by this bill—the 30 or so larger fillet and surimi catcher/processors that primarily target on pollock. In fact, the Bering Sea pollock fishery has been cited by the United Nations Food and Agricultural Organization (FAO) as the cleanest fishery in the world insofar as bycatch and discards are concerned. S. 1221 confuses the bycatch and discard of mid-water trawl pollock vessels with the bycatch of vessels that target on other groundfish species. It is these vessels, many of which are under 165 feet, that are responsible for most of the bycatch and discards that occur in the Bering Sea groundfish fisheries.

A further irony of S. 1221 lies in the fact that it blames wasteful fishing practices on the vessels that have actually been among the most responsible in terms of the fishing practices that they employ, the low bycatch rates that they achieve, their regulatory compliance and their support of progressive management measures in the North Pacific. The bill ignores the one company in the North Pacific that operates a fleet of bottom trawl vessels that has repeatedly ignored fishery manager's and coordinated industry efforts to control bycatch. Although this company seems to consistently conduct its operations without regard to bycatch rates or the effect that irresponsible fishing practices have on more conscientious fishermen, it would escape the more onerous aspects of S. 1221 since it appears to meet the proposed U.S. ownership standard.

SAFETY ISSUES

Even more important than economic considerations, however, are safety issues. Weather and sea conditions in the Bering Sea can be harsh, particularly in January and February when the pollock fishing season begins. From the standpoint of reducing risk in the fishing profession, large fishing vessels afford a relatively safe working environment. A review of U.S. Coast Guard statistics from the North Pacific indicate that only five lives were lost on factory trawlers larger than 165 feet between 1987–1997. During the same period, 260 lives were lost on fishing vessels less than 165 feet in length. In 1996, . . . national standards were revised to promote the safety of life and property at sea. It would be unwise to water down this important change in the law by passing legislation phasing out many of the safest vessels in the North Pacific fisheries.

The issue of regulating sizes of fishing vessels should be left to the regional fishery management councils. The characteristics of individual fisheries vary widely. In certain cases, there might be legitimate conservation, economic or social reasons to impose vessel size limits, but Congress ought not to impose a one-size-fits-all fisheries management rule.

> "If fish stocks were privately owned, incentives would exist to conserve them."

ESTABLISHING PRIVATE OWNERSHIP OF FISHING RIGHTS CAN PROMOTE SUSTAINABLE FISHING

Birgir Runolfsson

Many countries have attempted to stabilize fish populations in their areas of jurisdiction by calculating and enforcing a Total Allowable Catch (TAC) for certain fish species. Birgir Runolfsson maintains in the following viewpoint that such centralized regulation fails to address what he considers to be the root problem—the ocean's fish are nobody's property until they are caught. He proposes giving fishers property rights over portions of fish stocks. A possible method of achieving this, he argues, is a system of Individual Transferable Quotas (ITQs), under which companies and individuals are allocated a set percentage of the TAC. Holders are then free to use, buy, or sell their ITQs as they wish. Runolfsson contends that fishers will then have the incentive to manage fish populations and to police themselves against overfishing. Runolfsson is an economics professor and director of the Center for Rights-Based Fishing at the University of Iceland.

As you read, consider the following questions:

1. What have been the effects of government regulations on fishing, according to the author?
2. What four claims of ITQ critics does Runolfsson address?
3. What two countries that have established property-rights-based fisheries does Runolfsson examine?

Abridged from Birgir Runolfsson, "Fencing the Oceans," *Regulation*, vol. 20, no. 3, 1997. Reprinted by permission of the Cato Institute. *Endnotes in the original have been omitted in this reprint.*

Over the past several decades, countries have shifted the management of ocean fisheries within two hundred miles of their coastline from open access to intensive regulation. Governments attempt to restrict the total harvest of fish in order to stabilize or increase fish stocks. Yet regulatory regimes largely have failed to stem the decline of fisheries because they do not alter the fundamental incentives that lead to overfishing.

Change is inevitable in the fisheries. Retaining the status quo is not an option. Managing a fishery through regulation does not solve the basic incentive problems caused by the lack of property rights to the fish stock. Excessive fishing still exists because of the absence of property rights.

Recently, several countries have replaced fisheries managed by government with systems based on property rights. Rights-based fishing is increasingly recognized as a practical alternative to the inefficiencies of direct controls and regulation. On land, the conversion from medieval common ownership to the private property system is responsible for increases in economic productivity. The expansion of property rights as a method of economic organization should extend to individual transferable quotas in fisheries. As with property rights on land, the use of individual transferable quotas for fish will yield substantial economic benefits.

THE FISHERIES PROBLEM

Only a generation ago, the supply of fish available from the world's oceans seemed plentiful. However, advances in fishermen's ability to catch, preserve, transport, and sell fish quickly exceeded the ability of fish stocks to reproduce. Catches increased more than fourfold from 1950 to 1990, from twenty million metric tons to almost one hundred million metric tons. The United Nations Food and Agriculture Organization (FAO) maintained in 1993 that thirteen of seventeen major global fisheries were depleted or in serious decline. FAO also estimated that the world's fishing fleet catch was worth $72 billion but cost $92 billion to catch.

By the early 1980s, commercial fishing fleets had become so large and efficient that fish abundance and average catch per day for major stocks declined to a level that threatened stock reproduction. Many fisheries were unprofitable without subsidies. Although overall catch has remained constant in recent years, the increased catch of low-value species used for fishmeal has masked the decline of more commercially valuable species. Fish firms have responded to the decline of valuable species with the use of more capital and technology to increase the intensity of their fishing effort, exacerbating the decline of fish stocks.

Governments have responded to the decline in fish stocks with command-and-control regulation. Those regulatory regimes attempt to reduce overfishing through three types of restrictions: limits on the amount of time during which fishing can occur, limits on the types of capital and labor used to fish, and limits on the amount of fish caught. The length of fishing season, the size of the allowable catch, restricted fishing areas, number of fishermen, vessel size, and equipment, have all been regulated at various times.

While such regulations drive up costs and discourage some fishing effort, they do not alter the fact that fish are valuable but no one owns them. Those who catch fish earn money. That fundamental fact, as well as the existence of government subsidies in many countries—including the United States—for the acquisition of boats and gear, encourage fishermen to explore further means for finding fish. For example, limits on vessel size encourage investment in more boats and in more sophisticated equipment; specifying which days of the week, month, or year one can fish encourages more intensive effort on those days. Restrictions on fishing efforts make fishing less efficient than it could be. Seasonal closures coupled with improved fishing technology most often results in overcapitalization and wasteful racing for fish.

CREATING PROPERTY RIGHTS

Overfishing and other inefficient fishing practices have nothing to do with the nature of the resource, the characteristics of fishermen, or the localities in which fish are found. Rather, inefficiencies are the direct result of the definition and enforcement of property rights in fisheries. Fisheries are troubled by overfishing because they are not privately owned. Fishermen only own what they catch. The government, which is to say, everyone and therefore no one, owns the stock of fish from which the catch is taken.

If fish stocks were privately owned, incentives would exist to conserve them because the gains from their preservation as well as the costs of their exploitation would accrue to their owners. Private owners will neither race to take fish nor deplete stocks that would enhance future catch because if an owner does either, he bears the cost.

The establishment of private ownership in coastal fisheries, where fish stay put, is conceptually simple and very analogous to private property on land. A coastline could be carved up and private owners would be allowed to take exclusive possession of the fish in their area. Those rights, are called exclusive user rights (EURs) or territorial user rights in fisheries (TURFs). A single firm or fisher with EURs is assigned the right to a fishery within a country's jurisdiction. TURFs split the fishery within a country's jurisdiction into several geographic territories. Each territory is assigned to a single firm or a small group of fishers.

A Promising Solution

One of the root problems in U.S. fisheries is overcapitalization, the existence of excess harvesting and processing capacity in the fisheries. Overcapitalization threatens the economic and social fabric of the fishing industry. . . .

A promising solution to the problem of overcapitalization is a fishery management tool known as individual transferable quotas (ITQs). Under an ITQ plan, fishing privileges are assigned among participants in the fisheries. Each individual's quota share is expressed as a percentage of the total allowable catch in a fishery, and fishers have the flexibility to harvest their own quota share as weather or market conditions dictate. With the race for fish eliminated, there is no economic incentive to maintain capacity in a fishery beyond what is needed to harvest and process fish over the course of a fishing season. Also, because assigned quota shares are marketable commodities that can be leased, sold or otherwise transferred, new entrants are afforded an opportunity to participate in a fishery.

At-Sea Processors Association, "Individual Transferable Quota Programs," 1996.

EURs are appropriate for coastal fisheries in which the catch is small and involves only a single species. In Iceland, for example, the quahog fishery is organized with an exclusive user right. The fishery is small and a single vessel has a license for the fishery.

TURFs are appropriate for fisheries that are large and can be divided into geographic territories. A single individual or a small group can be assigned exclusive rights to a slice of an area where a species is located. The slice, or TURF, would usually be a rather small area close to shore. An example may be seen in the informal structure of the Maine lobster fishery.

Exclusive ownership of coastal fisheries would eliminate the need to regulate the fishery. The private owners of the fishery would have incentive to look after the maintenance of the coastal fish population. They would have the authority to prevent overfishing in their area.

The difficulty of defining boundaries and monitoring trespass in a liquid without obvious property lines emerges further away from the coast, where commercially valuable species of fish are found. The difficulty is as much a matter of incentives as technology. To be sure, new technological developments, such as remote observation by satellite, have enhanced the feasibility of assigning area rights to fishing grounds further offshore. But many other technologies no doubt already exist and are not recognized because incentives are lacking for their use.

INDIVIDUAL TRANSFERABLE QUOTAS

The solution for overfishing of migrating species is not as simple as the coastal situation solution. Most governments currently limit the Total Allowable Catch (TAC) in fisheries within their two hundred mile limits, though sometimes those limits are not strictly enforced. The problem with such limits is that if the fishery is simply closed once the TAC is reached, fishermen race against each other to get as large a share of the TAC as possible.

A system of Individual Transferable Quotas (ITQs) would modify simple TAC regulations to prevent that race. Such a system was instituted in Iceland in 1984 for all the major fisheries. Under an ITQ system, the TAC is allocated as individual quotas to fishermen, fishing firms, or fishing vessels. After the initial quotas are set, fishermen are free to adjust their share by buying, selling, or leasing a quota. That approach allows fishermen to better respond to market conditions by adjusting the nature, timing, and scale of operations to produce a more profitable harvest.

The quotas in an ITQ system should be proportional (the right to a percentage of the TAC) and permanent property rights. Absolute changes in the TAC will then translate into proportionate changes in each individual's quota holdings without any adjustment in the ITQ. The ITQ also should be allocated in perpetuity. Fisherman with a permanent interest in the harvest would manage their behavior more efficiently.

An ITQ system giving operators a right to a share of the harvest is not as good as a right to all fish in a defined territory. ITQs are not ideal because the gains from behavior that negatively affect the stock of fish, like cheating on one's quota, accrue only to one person while the losses are dissipated among all other owners of the quota. But because ITQs provide security over one's share of the harvest, fishermen will not dissipate the wealth in a fishery by competing among themselves for a greater share of the total catch. Even though ITQs are not ideal property rights, they provide a practical and politically achievable reform for existing ineffective systems of government fisheries administration.

CONCERNS ABOUT AN ITQ SYSTEM

Critics of an ITQ system make four claims. First, the understanding of fish stocks is insufficient to determine the correct TAC. Second, ITQ systems are more expensive to manage than traditional fish management systems. Third, ITQ systems exclude poor fisherman from their livelihood. And fourth, the government will regard quotas as simple property and thus, subject to a range of civil procedures such as seizure for bad debts or sale to settle a divorce. Such cavalier treatment of quotas is not compatible with sound management of a fishery.

The critics are correct that fisheries management is as much art as it is science. But the scientific limits of our knowledge of fishery dynamics affect the status quo and an ITQ system equally. That is because the TAC concept is a central feature of both. Even if TAC is not an explicit part of current politically managed systems, the implicit purpose of the restrictions and regulations in the status quo is to limit the catch to a level that a fishery can tolerate. And the explicit TAC in an ITQ system is preferable to the indirect ineffective methods of limiting the catch found in the status quo.

The benefits of an ITQ system exist even in the presence of scientific uncertainty about the long-run sustainability of any particular TAC. Continuous adjustment of TAC will be necessary because of the inherent biological variability in fisheries and their ecological interrelationships. Our understanding of those issues and hence, our ability to set TAC at a sustainable level should improve over time. Whether the TAC is set too high or too low will not affect the assertion that ITQs will maximize income from the TAC. For most fisheries, only a TAC that is set too high year after year will create difficulties.

ADMINISTERING ITQS

To be effective, any fisheries management scheme has to be monitored and enforced. One criticism of ITQs is that such schemes are more expensive to administer and enforce than traditional types of schemes. All fisheries management schemes have costs. The advantage of ITQs is that they focus attention on the explicit costs of management versus the economic benefits. Improvements to management are more likely to be initiated if the costs of management are transparent.

Monitoring and enforcement need not be a government responsibility. Indeed, there is considerable scope for self-policing in a fishery. Large numbers of fishers spend time on the water harvesting their catch. They can and will be enlisted in policing the resource. The incentive for self-policing follows directly from the ownership of quota. Although individuals profit if they exceed their quota (steal fish), it costs them if other quota owners do likewise. If everyone exceeds their quota, the fishery will be overfished, fishers income will fall, and the price of quota will fall. Fishermen themselves will, in time, protect their property rights just like landowners protect theirs.

The perception that closing the commons excludes some from access to fishing is true but the concern is overstated. The fishing of ocean resources is currently excessive, so by definition, some who are currently fishing will not be fishing in the future. But that fact is unaffected by the management system in place. The ITQ system, in fact, is superior to the traditional system because as long as people can trade the quota rights, nobody is automatically excluded. And once you obtain an ITQ right, the fish will actually exist for you to catch. Under a traditional system, everyone is free to fish, but the race to harvest often implies that no one is entitled to a fish. . . .

In some cases, the argument that ITQs allow the use of fisheries by some people to the exclusion of others is nothing more than an argument against the institution of private property. The long and bitter experience with public ownership of resources in Eastern Europe suggests that the argument should be put the other way; lack of private ownership allows the exploitation of resources by some to the detriment of others.

DISTRIBUTING QUOTA RIGHTS

In contrast, a legitimate concern in the creation of an ITQ system is the mechanism used to distribute the initial quota rights. An auction favors those who have access to capital. A lottery favors those who are lucky. Allocation to existing fisherman favors history.

Two important economic truths should govern any discussion of the initial distribution of quota rights. First, the initial distribution of quota does not affect efficiency; as long as quotas are easily traded, those who can use them most efficiently will purchase them. Second, the concerns of those who worry about the exclusion of some from the new system can be ameliorated in the design of the initial quota distribution system. For example, if "little" fishermen are a source of concern, give "little" fishermen more initial quotas then they would receive if quotas were initially distributed according to historical catch data. Hence they can either sell fish or sell the quotas to larger firms and invest the proceeds of the sale to raise their incomes.

The final concern of the critics is also true but irrelevant. Some worry that because ITQs will be considered the property of fishers, the government or courts will seize ITQs to satisfy debts, lawsuits, or other judgements against a quota holder. As with other classes of assets, quotas would be split in divorce settlements and inherited as a quota owner passes away.

That charge is true but not an argument against private ownership of fisheries. If a quota owner runs up debts, he may be obliged to sell his quota. But the new owner also will have the same incentive to manage his asset competently. Improved management of the fishery requires exclusive rights, and exclusive rights require that someone be responsible. Responsibility implies the possibility of asset loss.

EXAMPLES OF RIGHTS-BASED FISHERIES

Although no country has yet completely privatized its fisheries, many countries have experimented with property-rights-based management including Australia, Canada, the Netherlands, Norway, Portugal, the United Kingdom, South Africa, and the United States. But New Zealand and Icelandic fisheries offer the best examples since both have used property rights management more extensively than other countries.

New Zealand has developed the most extensive property-rights-based fisheries in the world. The Ministry of Fisheries, after consulting with scientists and industry representatives, sets TACs for the commercial species in each fishing area within the New Zealand jurisdiction. The quotas are permanent, perfectly divisible, and transferable, but no owner may own more than 35 percent of total deep-sea quota and 20 percent of total inshore quota.

The most important innovation of the New Zealand experience is the quota management company. It was formed by the quota owners in the same manner condominium owners in a large building form a management company to oversee their collective interests. The companies can potentially solve, through contract negotiation, any discrepancies between the interests of individual quota owners and the interests of the fishery as a whole. The New Zealand ITQ system was initially set up without any means for enabling quota owners to act collectively. In spite of the fact that they still lack the legal right to manage or enhance their fisheries, quota owners have organized themselves into management companies.

For example, in the deep-water orange roughy fishery, quota owners have formed a joint management company, the Exploratory Fishing Company, to undertake exploratory research into orange roughy fish stocks and facilitate other management strategies. The Challenger Scallop Enhancement Company conducts research in the scallop fishery, implements its own compliance regime, and develops its own management plans in conjunction with other users. The improvement to the fishery is exceeding expectations.

The ITQ system experience is favorable. The exclusive right to harvest the resource guaranteed by the ITQ system has impelled New Zealand fishermen to treat fisheries as an asset. Overall the change has been from a system of short-term to a system of long-term fisheries management. Aggregate catches have increased and most resource stocks seem to be stable. In 1996 the TACs for twenty-nine of thirty-two ITQ species exceeded their 1986 TACs. Harvest quality has improved and there is evidence of reduced fishing effort. The size of the fishing fleet has declined slightly since 1990 and catch per unit effort has been stable or slightly increasing. Profitability in the industry has been good and improving. Both industry and government are generally satisfied with the system.

ICELAND'S FISHERIES

Second to New Zealand, Iceland has developed property-rights-based fisheries most extensively. Icelanders introduced individual vessel quotas (IQs) in 1975 in the herring fishery. Those quotas applied to catches by individual vessels. They were similar to ITQs except they were not transferable. In 1979, the IQ regime was transformed into an ITQ system for that fishery. In 1990, all fisheries became subject to a comprehensive ITQ system, with only minor exemptions. The ITQ system is a proportional or share quota system. The number of species under the ITQ system has increased to thirteen, from five in 1984.

The Ministry of Fisheries, on the recommendation from Iceland's Marine Research Institute (MRI), an independent government institute that conducts oceanographic and fisheries research, recommends TACs for all commercial species. The basic property right in the system is a share of the TAC for every species for which there is a TAC. The quotas are permanent, perfectly divisible, and transferable. There are no rules of maximum quota holding. . . .

The Icelandic ITQ system was created due to sharply declining stocks of herring and cod. The experience with the ITQ system is generally favorable. Catches of herring have increased. And more importantly, catch per unit effort has increased significantly, for example, by more than tenfold in the Icelandic Herring fishery. In fact the condition of the herring stock is better than at any time since the 1950s. The number of vessels in the fishery has declined from more than two hundred in 1980 to less than thirty in 1995, although the average vessel size has increased substantially.

The groundfish fisheries, for example cod fisheries, have not improved as much because the TACs have been set too high. Politicians have chosen the gradual approach to cutting the cod catch, despite recommendations by the Marine Research Institute. Only recently has the TAC been set in accordance with MRI recommendations. That was done at the insistence of the Association of Vessel Owners who want to preserve the value of their ITQ assets. Stocks seem to be rebounding as a result; the current TAC in the cod fishery is 20 percent higher than last year. The fact that the government had to respond to pressure from quota owners to protect the value of their property demonstrates the dynamics set up by the ITQ system.

Since 1990, when the comprehensive ITQ system came into effect, there have been substantial improvements in cod fishery economics. Fishing effort is now more than 30 percent lower than it was in 1983. Fishing capital, which had increased by more than 400 percent in 1960–1990, has actually declined since 1990, and the number of vessels has also declined. Harvest quality and profits have improved significantly and fishing effort has been reduced.

TOWARDS PRIVATE PROPERTY FISHERIES

ITQ systems could prove to be one of the great institutional changes in recent history. Enclosure and privatization of ocean resources could be comparable to the land enclosure movement in British history or the fencing of western range land in American history. As with the enclosure of common land resources, the establishment of property rights in fisheries conserves the resource. Both land enclosures and ITQs remove the threat of overexploitation of resources that results from open access.

Under an ITQ fisheries scheme, quota owners form, in effect, a club with the exclusive right to harvest fish species commercially. Their property right in the fishery, in the form of their ITQ, will reflect the overall value of the fishery. The situation is exactly analogous to property rights on land. If a property owner does not maintain and improve his or her property, then its value will fall. If the property is well maintained, its value will rise. Club members' wealth increases if they encourage fisheries management strategies that improve the health of the fishery.

"The verdict is still out as to what extent ITQs confer conservation dividends."

ALTERNATIVES TO PRIVATE OWNERSHIP CAN BETTER PROMOTE SUSTAINABLE FISHING

Anne Platt McGinn

In the following viewpoint, Anne Platt McGinn argues that the fishing industry must be transformed from a heavily capitalized and subsidized commercial industry into one that is more sustainable, community-based, and environmentally aware. The measures McGinn prescribes include setting up community-based planning groups to manage coastal resources, encouraging part-time or seasonal fishing, establishing regulations to limit the capacity and access of fishers, and educating consumers to support sustainable fishing. She maintains that Individual Transferable Quotas (ITQs), a system of privatizing fishing property rights advocated by some observers, tend to concentrate fishing rights into the hands of those that can afford them and may do little to prevent overfishing without reliance on other regulatory measures. McGinn is a research associate with the Worldwatch Institute, an environmental research organization.

As you read, consider the following questions:

1. Why should fishers be included in plans to manage fisheries, according to McGinn?
2. What happened to the fishing industry in New Zealand after the introduction of ITQs, according to the author?
3. What impact does McGinn believe consumers can have on the fishing industry?

Reprinted from Anne Platt McGinn, "Rocking the Boat: Conserving Fisheries and Protecting Jobs," Worldwatch Paper 142, June 1998, by permission of the Worldwatch Institute.

Fish and fishers would benefit greatly if the urge to divide up the world's oceans and to control fisheries were tempered with ecological and social realities. Given the deteriorating condition of marine resources, fishers need to move quickly away from today's overcapitalized, heavily subsidized industry to a more sustainable, biologically sensitive, and socially diverse one. This transition will spark some short-term economic and social disruption, but will allow fisheries to recover and fishers to enjoy greater returns in the long run. A key issue is the extent to which society will guide these changes and help ease the short-term economic hardships—or merely let the chips fall where they may.

One thing is clear: It is no longer enough for fishers simply to know how to operate a boat and catch fish. They need to be intimately involved in managing and conserving the resource. Whether the day-to-day tasks and responsibilities of fisheries management are devolved to the community level, shared between fishers and local or national governments (comanagement), relegated exclusively to private interests, or turned over to the state—or some combination of the above—bringing fishers together with scientists and resource planners is key to reconciling ecological realities with economic pressures to mine fisheries.

Changes on two other broad fronts are also important prerequisites to sustainable fisheries management. First, it is critical to bring conservation back to the forefront of fisheries management. A more holistic, ecologically based approach that takes into account where fish live and where land-based activities take the greatest toll will go a long way toward renewing fishery resources. Second, economic measures and policy changes are necessary to reduce capacity and limit access. An obvious policy to achieve this aim is for governments to tackle the issue of reducing subsidies for fishing fleets and to overhaul economic incentives that bolster fishing capacity.

INTEGRATED COASTAL MANAGEMENT

To promote conservation, fishers and officials need to view fish as part of a larger ecological system, rather than simply as a commodity to extract. One tool that can help bring about this change is integrated coastal management (ICM). Through the process of community-based planning, ICM brings together diverse groups of people—fishers, politicians, tourism operators, traders, and the general public—to identify their shared problems and goals, and to define solutions that build on their common interests. Discussions, mapping exercises, and site visits all help people make the connections between land and water use, and the health of fisheries.

More than 10 years of experience with integrated coastal management in the Philippines has revealed the importance of approaching fisheries conservation as a series of steps in an ongoing process. To this end, it is important for fishers—and the community at large—to witness firsthand the connections between restored habitat and improved fish catches. Mangrove replanting projects and construction of artificial reefs are two concrete steps that help some fish stocks rebound quickly. Once people appreciate the immediate results of their work, they are more likely to engage in longer-term protection efforts, such as marine sanctuaries, which involve removing an area from use entirely.

"No-take" fishing areas, seasonal fisheries closures, and marine protected areas also help depleted stocks rebound—and profits return—by limiting accessibility and easing pressures on the resource. A 1982 study of the offshore shrimp fishery in Texas waters documented a $9 million increase in the value of the shrimp catch following the closure of that fishery, which allowed juvenile shrimp to grow to more marketable sizes. In 1997, a group of international marine scientists called for governments to increase protected marine reserves from the current 0.25 percent of the ocean's surface area to 20 percent by 2020. Such a move would provide a much needed respite for commercially depleted fisheries and, combined with effective management of areas that continue to be fished, could allow global catches to grow by an estimated 10 million tons.

INDIVIDUAL TRANSFERABLE QUOTAS

In addition to habitat protection, fisheries conservation would benefit from policies and tools that help to restrict use of the resources—whether through marine protected areas or closed fishing seasons that apply to all fishers, or through more limited forms of entry such as Individual Transferable Quotas (ITQs). In an open access or shared resource, ownership does not materialize until fish are in the boat. ITQs, on the other hand, seek to extend ownership to fish while they are still in the water. However it is achieved, restricted use is no guarantee that fish will return. But if protective measures are imposed in an equitable and careful manner, people will more likely abide by the restrictions, thus dramatically improving the chances of recovery. Furthermore, reducing immediate pressures on the resource does buy time for fishers and policymakers to implement more drastic—and needed—cutbacks that will yield greater returns in the long run.

Many governments are moving toward market-based systems

for limiting access, of which ITQs are the most publicized. ITQs essentially privatize the right to fish by conferring on fishers a quota for a certain share of fish in a particular area and time which can be bought and sold on the market. Since individual shares rise or fall in proportion to changes in the total allowed catch, the theory is that fishers will go to greater lengths to protect the total stock, and thereby enjoy greater individual returns.

ITQs No Solution

Individual Transferable Quotas, or ITQs, would privatize fisheries by allocating exclusive ownership rights to a percentage of the harvest to boat owners on the basis of prior catch history. . . .

ITQs are promoted as a panacea for the ills of overcapitalization and overfishing. Proponents claim that ITQs are a way to limit the number of boats, extend shortened fishing seasons, reduce waste and bycatch, and improve safety. However, small-boat owners and fishermen using cleaner fishing gear would be excluded from most fisheries based on their smaller financial means and lower percentage of the catch. In the North Pacific cod fishery, for example, factory trawlers would receive ownership rights to over half the total allowable catch, based on the recent history of the fishery. This despite the fact that factory trawler bycatch rates in the cod fishery are among the highest of any fishery—roughly 40% of the catch is non-target species, including halibut and crab. Boat owners using the cleaner pot and jig gear have only a 2% bycatch rate, yet they would receive ownership of only a few percent of the fishing quotas under an ITQ system. ITQs are not a solution to the problems of overfishing.

Greenpeace, *Sinking Fast: How Factory Trawlers Are Destroying U.S. Fisheries and Marine Ecosystems*, 1996.

Now implemented to varying degrees in Australia, Canada, Iceland, the Netherlands, New Zealand, and the United States, ITQs reduce overcapacity by enabling inefficient fishers to sell their quotas and thereby receive compensation for withdrawal from the industry. But this system of compensation has tended to concentrate quotas, and therefore access rights, in the hands of the few who can afford them. Just one year after the first ITQ program in the world was implemented in 1986 in New Zealand, the country's three largest fishing companies held title to 43 percent of the ITQs. Within five years, these three companies controlled half of all ITQs in the system. Many small-vessel owners and fishers either subcontracted to larger companies or went out of business completely.

Modifications have since been applied in New Zealand, and more recent programs in other areas have attempted to address the social impacts of ITQs. Maori fishers in New Zealand now receive set shares of quotas, whereas Alaska has developed Community Development Quotas for community-based, small-scale fishers. In Alaska, New Zealand, and other places, policymakers have set limits on consolidation of ownership to prevent monopolies from forming.

The verdict is still out as to what extent ITQs confer conservation dividends. Although ITQs do tend to reduce pressure on individual fishing grounds and species, they impel fishers to move elsewhere. Quotas also encourage fishers to select only the highest-quality species, thereby increasing bycatch. Of the 31 ITQ fisheries reviewed by the Committee for Fisheries of the Organisation for Economic Co-operation and Development (OECD), most used other regulations, such as catch-size limits, gear restrictions, and closed seasons, to help mitigate the shortcomings of quota systems.

CHARGING FISHERS

A better way to protect fish may be to compensate governments directly for extraction of natural wealth. Charging fishers based on how much they take would also help pay for some of the costs of managing them. Many industrial countries now charge a flat fee for fishing permits and licenses; levying a tax on a particular gear type or practice would be a logical extension of this practice. Indeed, Australia now charges its domestic fishers about 2.5 percent of the value of landings earned at the dock. Worldwide, governments forgo between $3 and $7 billion in uncollected user fees from domestic and foreign fishers, according to estimates by Matteo Milazzo.

In addition to charging fishers for the privilege of extracting the resource, reducing subsidies would also help protect fish stocks—and save governments and taxpayers money in the process. Economic programs that were originally intended to help fishers have helped destroy the resource by prolonging the notion that fishing is free. It is particularly important to engage the governments of China, the European Union, Japan, Norway, Russia, and the United States in a global subsidy reduction scheme because of their central role in providing subsidies. In contrast to the billions of dollars they spend propping up bloated, inefficient fleets, only about $500 million—5 percent of total subsidies from these governments—is budgeted to reduce fishing capacity.

EMPLOYMENT ISSUES

Of course, reducing overcapacity leads to questions about employment issues and what will happen to the people who rely on fishing for their jobs. In overcrowded industrialized fisheries, a combination of short-term economic aid to help excess laborers move out of the industry, and longer-term restructuring along with overall downsizing in the industry, can help solve this problem.

In industrial countries, governments have paid to retrain fishers for other jobs or to buy back vessels, often with little success because they failed to implement the other crucial step: reducing the 95 percent of subsidies that motivate people to stay put and even enhance their capabilities. A nearly $2 billion five-year social adjustment program in Atlantic Canada is helping fishers stay out of debt for the time being, but critics argue that it amounts to nothing more than a massive social welfare program. In 1996, the European Union set aside $2.2 billion to pay for job-retraining programs and economic aid for out-of-work fishers. But intense pressure from industry officials, politicians, and fishers prompted EU ministers to backslide on fleet reductions and to postpone needed reforms.

The challenge of finding work for displaced fishers is more difficult in developing countries because more people there rely on fishing for their livelihoods, alternative jobs are more limited than in industrial nations, and governments have scarce financial resources to devote to the problem. Also, in coastal communities from Atlantic Canada to Southeast Asia, fishing is not just a job, it is a way of life. In a survey of nine community-based coastal resource management projects in the Philippines, more than 80 percent of fishers said they would not leave the sector, even if they had a job with comparable income, because their lives and culture were so closely linked to the sea. Rather than trying to create jobs for people who want to stay put, some authorities are drawing on the tradition of part-time or seasonal involvement in fishing, with supplemental income from other activities. The result is that overall pressure on the fish is still reduced, but the wealth is spread over a larger number of people.

REGULATING WHERE AND HOW PEOPLE FISH

Other communities hope to preserve fish as an important source of food and jobs by keeping large boats out of inshore and coastal areas where small-scale fishers tend to work. With nearly 30,000 islands covering an area greater than the continental United States, Indonesia has excluded medium- and large-scale fishers from an inshore zone since 1980. While intrusions into this zone do occur, the ban has protected an important way of life and has encouraged indigenous fishers to become more involved in patrolling their waters. Similarly, the more than 7 million small-scale fishers operating in India's coastal areas successfully pressured their government to prohibit joint ventures with foreign vessels in 1995, and the Supreme Court of Chile banned U.S. fishing trawlers from entering their waters in late 1997.

Instead of regulating where people fish, some countries are now restricting how they fish. Namibia, for example, has enacted a no-discards law that prohibits throwing edible and marketable fish back to sea. The regulations apply across all sectors of the fishery—most of which operate under a quota system—and are enforced by onboard observers. Since implementation in the early 1990s, the policy has received widespread industry support and is viewed as highly effective at reducing discards. Norway passed similar legislation in 1996.

U.S. fishery officials are in the process of writing tougher by-catch laws that require the use of more selective gear. Turtle excluder devices (TEDs) costing $75–$500 apiece are required on trawl nets used to capture shrimp under the U.S. Endangered Species Act. Shaped like a funnel in the net that allows turtles to escape drowning, these devices have helped preserve the last Kemp's Ridley turtles in the Gulf of Mexico, as well as numerous finfish. Worldwide, investments in more selective gear types could yield reductions in discards of nearly 60 percent, resulting in an additional 15 million tons of fish landings per year, according to the FAO.

Another issue that continues to undermine fisheries worldwide is the lack of enforcement and monitoring of existing agreements and bans. In the tropics, many communities are now working to educate fishers and the public about the dangers of cyanide fishing and to generate support for enforcing a ban on its use. For deep sea fisheries that are far removed from public oversight, Australia and New Zealand use satellite-based systems to monitor vessel movement and to make sure they do not move into no-fishing areas or grounds limited to national fishers. The South Pacific Forum Fisheries Agency (SPFFA) is planning to adopt a similar system to monitor the position of hundreds of vessels in South Pacific waters.

THE NEED FOR DATA

Ultimately, the success of all the management tools depends on accurate fish population data and conservative catch limits that take biological uncertainty into account. Movement in this direction, however, is hampered by the fact that most governments in developing countries lack a ministry of fisheries. Consequently, fishery issues come under the purview of the agriculture or commerce ministry, which does not consult the scientists and resource users who know the fish stocks and appreciate the inherent variability in a large-scale and complex ecosystem.

One small but significant approach to addressing the need for better data comes from coral reef scientists and coastal managers, who have recently enlisted the help of recreational scuba divers. Numbering more than 7 million worldwide, sport divers who volunteer to collect data are given basic training to identify and survey fish and coral species, and to conduct rudimentary site assessments. The data are then compiled by scientists and put into a global inventory available on CD-ROM, known as ReefBase, which policymakers use to monitor the trends and conditions of reefs and to target intervention and protection programs. More efforts that engage the help of concerned individuals and volunteers would help overcome funding and data deficiencies and build greater public awareness of the problems plaguing world fisheries.

In addition to on-site surveys, determining the effects of wide-scale changes in oceanographic conditions is a relatively new pursuit that deserves more attention. For example, more accurate models of ocean temperature, salinity, and currents are now being developed that can help officials better distinguish between changes due to overfishing and those due to natural stock migration or changing ocean and climate conditions. Officials can then adjust catch limits accordingly.

THE ROLE OF CONSUMERS

Promoting sustainable fisheries also means addressing the growing demand for fish. To shift demand away from environmentally damaging fishing techniques and products, market forces, such as charging consumers a higher price for such selections, can be enlisted. A U.S. consumer boycott of canned tuna in the late 1980s successfully forced changes in the way fishers caught tuna and helped protect dolphins from being ensnared in purse seine nets. In April 1996, the World Wide Fund for Nature teamed up with one of the world's largest manufacturers of seafood products, Anglo-Dutch Unilever, to create economic incentives for sustainable fishing. Implemented through an independent Marine Stewardship Council, fisheries products that are harvested in a sustainable manner will qualify for an ecolabel.

More efforts like these could help persuade manufacturers to replace fish oils and fishmeals with other products, to convince fishers to curb wasteful practices, and to generate greater public awareness of the need to carefully choose what fish species and products to eat.

Consumer education can also tackle the growing demand for live fish and specialty products that result in high environmental costs. People who enjoy shark fin soup, live fish prepared before their eyes, and other aquatic delicacies can be charged the full ecological price for their meals, for it is their demand that drives the use of deadly practices such as cyanide fishing.

Ultimately, what determines the shape that fishing will take in the future begins at home—far from the high seas and underwater habitats that are under siege. Consumers can have a positive impact on global fisheries by buying fish products that have been made sustainably, asking where a fish came from and how it was captured or raised, and demanding that policymakers support the recommendations of scientists to close fisheries and reduce the amount of fishing.

The current habit of reacting to fishery crises rather than preventing biological declines cannot continue indefinitely. With fewer and fewer commercially viable species available, fishers will soon face catastrophic losses from which few governments will be able to rescue them. By shifting our focus from what is done to fish to what can be done for fish, fisheries can continue to provide food, jobs, and enjoyment for hundreds of millions of people worldwide. Ultimately, such a shift can bring world fisheries back within the bounds of nature and allow this renewable resource to flourish.

"With human populations growing [and] wild fish resources shrinking, . . . aquaculture is the way to fill the gap."

AQUACULTURE CAN RELIEVE PRESSURE ON WILD FISH POPULATIONS

Christine Weber Hale

Aquaculture, which literally means "farming in water," refers to the domestication and raising of fish and other marine life. Most successful with freshwater marine life such as catfish and trout, aquaculture has become an increasingly popular source for some forms of seafood as well, including shrimp and salmon. In the following viewpoint, Christine Weber Hale examines the developing aquaculture industry, focusing on the western United States. She asserts that aquaculture can fill the gap between growing consumer demand for seafood and declining wild fish resources. Hale is a food writer for *Sunset* magazine.

As you read, consider the following questions:

1. What kinds of ocean life are being raised in aquaculture facilities, according to Hale?
2. What two basic types of aquaculture can be distinguished, according to the author?
3. What does Hale list as some of the criticisms directed at the aquaculture industry?

Reprinted from Christine Weber Hale, "Who Grew That Fish on Your Plate?" *Sunset*, August 1995, by permission of Lane Publishing.

It's 6:17 A.M. Bill Williams zips up his windbreaker, pulls his baseball cap down a little lower over his eyes, and steps into a blustering wind. Williams's crop needs his attention. He and his crew must harvest the day's orders before the tide retreats and strands his barge.

Tide? Barge? Williams grows Pacific oysters, and Williams's Shellfish Farms covers about 760 intertidal acres in Morro Bay, California. After daily harvesting, Williams's oysters are iced and shipped to such places as Galley Restaurant in Morro Bay and Finicky Fish Market and Fat Cats in Port San Luis. There, Williams's oysters join salmon from Washington and British Columbia, trout from Idaho, and, from California, Mediterranean bay mussels from Carlsbad and Santa Barbara, tilapia and hybrid striped bass from Palm Springs, and sturgeon produced near Sacramento. Bill Williams is just one example in a new and expanding breed of Western farmers who practice aquaculture.

Aquaculture began in the United States in the last century with oyster and trout farming, but Asian countries have farmed fish for thousands of years. What's new is aquaculture's skyrocketing growth. It is now the fastest-growing sector of U.S. mainstream agriculture. In 1994, 12 percent of all seafood consumed in this country had been domestically farmed. Experts predict that the percentage will rise to 25 during the next five years.

A phenomenal jump in global seafood demand at a time of declining wild harvests—down 7 percent in the last three years since 1992—has galvanized fish farming. World demand is expected to increase by 70 percent in the next 35 years. With human populations growing, wild fish resources shrinking, and nutritionists telling us to eat more fish, aquaculture is the way to fill the gap.

What does aquaculture mean to consumers? You can buy farmed fish and shellfish from all over the world, and your choices are no longer limited by the season or concerns about conservation. In the [American] West, abalone, catfish, clams, crayfish, mussels, oysters, salmon, hybrid striped bass, sturgeon, tilapia, and trout are farmed. Cultivated products such as blue lobsters and caviar are in the works. The safety and quality of farmed fish can be monitored. But even as aquaculture is reshaping future resources, it is one of the most hotly debated issues in the food industry.

WHAT IS AQUACULTURE?

Aquaculture, which simply means cultivating in water, can be divided into two basic types: extensive and intensive. Extensive aquaculturists "plant" their "crops" in a natural environment and let them develop on their own until they reach harvest-size. (Saltwater aquaculture is often called mariculture, indicating that the species are grown in a bay or the ocean instead of in manmade freshwater tanks or ponds.) Mollusks, specifically oysters, mussels, and clams, are mostly farmed in this way.

Intensive aquaculturists create a completely controlled environment. Often, water is pumped in and must be constantly cleaned as it recirculates. Machines dispense food at regular intervals—a full-time job if you have a million fish.

GLOBAL AQUACULTURE PRODUCTION

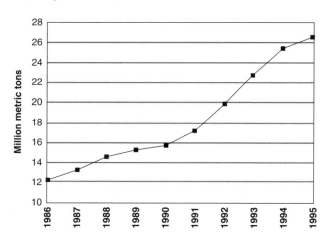

UN Food and Agriculture Association, *Review of the State of World Aquaculture*, 1997.

Abalone, some crayfish, hybrid striped bass, sturgeon, tilapia, and trout are raised in densely populated tanks (sometimes called raceways if they're long and narrow). Water moves continuously through these tanks. As fish and shellfish grow, they are kept in same-size batches to reach market-size simultaneously.

Some intensive farms utilize technology almost as complex as that used in space exploration. Solar AquaFarms in Sun City, California, for example, raises tilapia in a big way—5 million pounds annually. And it does it smack-dab in the middle of the desert. The farm, covered by huge greenhouses, sprawls across an area equal to 26 football fields. Inside the greenhouses, enormous solar-heated tanks create the steamy, tropical environment in which tilapia thrive—not unlike the Nile of ancient Egypt, where these fish are thought to have evolved. Each 60- by 500-

foot raceway holds half a million gallons and about 200,000 fish. Efficiency is vital. Every drop of water must be reused. A patented treatment and recycling system purifies the water and converts fish waste to organic fertilizers. Computers monitor and control feeding, water temperature, and oxygen and nitrogen levels; they also keep growth and population statistics. And if the computers fail, attendants are on-site 24 hours a day, ready to take over.

MARKETS AND MENUS ARE CHANGING

Tilapia comes from Solar AquaFarms 365 days of the year. Salmon floods the market—430 million pounds are farmed annually worldwide. Five years ago in 1990, you might have paid as much as $14.99 a pound for salmon. Today in 1995 farmed specials at $2.99 a pound aren't unusual. The old adage about avoiding months with the letter r no longer applies to shellfish safety. Sturgeon and striped bass, disappearing in the wild, are enjoying a revival because of aquaculture.

The food service industry values the convenience, steady supply, and relatively constant price that aquaculture brings to fish. Restaurants can plan menus with assurance that fish and shellfish will arrive. Some farms even offer customized fish. Kent SeaFarms, near Palm Springs, sells its hybrid striped bass in 10 sizes. Chefs praise the consistent quality and portion sizes of farmed fish, which help them standardize preparation.

But what if you're a consumer outside the food industry? How do you know if fish at a restaurant or at the market is farmed? Even if you're curious, don't expect to find signs. Restaurants and retailers are not required to label fish or shellfish as farmed—except in Washington, where a 1992 state law requires that farmed salmon be labeled.

Wild abalone, salmon, and sturgeon are seasonal. So when you see these fish featured as seasonal specialties, they are usually wild.

Even if fish is labeled as farmed, this information is frequently ignored when the fish is marketed through wholesale brokers. The salesperson or server may have no clue as to the source. Mollusks, however, are certified as to where and when they were harvested, and the seller must make the information available if you ask.

IS IT SAFE?

The safety of fish and shellfish, whether farmed or wild, is the biggest concern of both consumers and people in the aquaculture and fishing industries. Water quality often determines safety. Since wild fish swim where they want, waters they visit can't often be monitored. And even waters that are monitored aren't always places you'd want your dinner to hail from.

But farmed fish are another matter. Commercially grown clams, mussels, and oysters are raised in waters whose safety must be tested under the Food and Drug Administration's (FDA's) National Shellfish Sanitation Program (NSSP), administered by state health departments. Under NSSP, farmers must sample water from several sites in the harvest area each month, testing for biotoxins and bacteria. Shellfish are tested for natural biotoxins such as paralytic shellfish poisoning and domoic acid.

TAKING PRESSURE OFF WILD FISH SPECIES

At a time when many of the world's wild marine environments are in decline, aquaculture is trying to ensure that the next time you want to grill a fish or load a fork with shrimp, you'll have a reliable supply of healthy seafood. . . .

Aquaculture, which now provides between 15 and 20 percent of the US seafood supply, offers a bonus in humankind's quest for nourishment: It not only complements traditionally harvested seafood but also, by taking pressure off wild species, helps us buy time to figure out ways to restore the world's ocean and freshwater fisheries.

Dick Victory, *Washingtonian*, April 1997.

Only mollusks, often consumed whole and raw, are federally regulated as they grow because they are filter feeders—they ingest algae filtered from the water. If harmful bacteria are present, they concentrate in the meat and can cause illness or worse. Because other fish and shellfish aren't filter feeders, and don't ingest the toxins directly, they are slower to accumulate harmful bacteria. Monthly water checks around farmed mollusks spot pollution well before it reaches dangerous levels. If water and shellfish tests don't meet NSSP standards, the shellfish cannot legally be harvested or sold.

But no farmer wants to get to that point. John Davis, owner of Carlsbad AquaFarms in Carlsbad, California, raises Mediterranean bay mussels. He speaks for many aquaculturists when he says, "The last thing in the world I want is for someone to get sick from my product." Because Davis's mussels grow in a restricted area—the FDA has designated the water quality unpredictable—they must be cleansed before being sold. Davis has developed and built his own "depuration" system. After harvest, his mussels get a spa treatment, reposing as long as 48 hours in trays of purified water, which purges their meat of any impurities.

The FDA works with the state to inspect the handling, processing, and packing of all shellfish, including mollusks, and fish—but only sporadically unless there is a consumer problem. Within a year, the FDA hopes to introduce a new mandatory regulatory program, Hazard Analysis and Critical Control Point (HACCP). [Editor's note: HACCP policies were introduced in January 1998.]

Under the program, inspection focus shifts from after-the-fact correction to problem prevention. Fish processors will be required to police their own sites through a detailed system of record keeping and safety checks. The FDA will review the records and continue conducting on-site inspections. However, an FDA spokesperson says the number of inspectors will drop because of government downsizing; some people question how this change will affect the program, particularly as penalties for violations have yet to be established.

HACCP rules will apply to all domestic seafood processors, distributors, and importers, and also to foreign processors that export fish and shellfish into the United States. Retail sales and food preparation are not covered by the program.

CRITICS OF AQUACULTURE

Aquaculture's most vocal critics include commercial fishermen. For years, fishermen and fish farmers have traded insults and accusations—some legitimate, many unsubstantiated—in something of a range war. Farmers have been accused of dosing fish with antibiotics and steroids, and of producing mushier, less flavorful fish. Fishermen, in turn, have been blamed for depleting wild stock, and for careless and rough handling of their catches.

Although the infighting has undermined consumer confidence, it has shaken up questionable fishing and fish-farming practices. Now the collective chorus is a variation on Rodgers and Hammerstein's *Oklahoma!*: "The farmer and the cowman should be friends." The California Seafood Council works, when feasible, with the state Fish and Game Department and with individual fisheries in assessing and replenishing wild stock. And the California Aquaculture Association and California Seafood Council took another cooperative step when they signed a joint marketing agreement to promote California sturgeon.

But opposition to aquaculture goes beyond economics. For some, it's aesthetics. Saltwater farms along the coast are usually in sheltered waters or bays, and homeowners with views of the water generally don't want to look at fish farms.

Other critics have environmental concerns. Tom Worthington and Paul Johnson, owners of the San Francisco–based Monterey Fish company, acknowledge that aquaculture has increased the variety of fish they can sell year-round, but worry that this convenience masks a larger issue. Johnson says, "The constant supply that aquaculture provides blinds people to problems in the environment, such as the decline in wild fish populations." Worthington and Johnson also voice a fear, held by many, that genetically manipulated hybrids might escape and breed with their wild counterparts, leading to a decline in true wild species.

A BRIGHT FUTURE

Controversy aside, aquaculture is having its day in the sun. The greatest expansion is expected where land, labor, and water cost less, in South America and Mexico.

Western universities, marine laboratories, and other experimental sites are on the fast track to finding ways to farm more kinds of fish. And geneticists are researching ways to help fish mature faster. With so many breakthroughs, it now seems that the only factor limiting aquaculture is the same one hampering wild fish—the need for clean water.

| "One of the enduring myths of aquaculture—that farming cod or salmon will 'take pressure off the oceans'—is simply false."

AQUACULTURE MAY NOT RELIEVE PRESSURE ON WILD FISH POPULATIONS

Bob Holmes

Aquaculture—the domestic cultivating of marine life—has been touted as an alternative method of supplying the world with seafood. In the following viewpoint, Bob Holmes argues that the belief that aquaculture can replace commercial fishing and thus help stabilize wild fish populations is erroneous. Aquaculture has too many disadvantages, Holmes contends, to become a viable replacement for fishing. Chief among Holmes's concerns are the continued need of commercially caught fish to provide food for the farm-raised fish and the environmental damage aquaculture farms have caused in the places they have been located. Holmes is a writer for *New Scientist*, a British weekly magazine.

As you read, consider the following questions:

1. According to Holmes, why is tilapia a popular choice of fish for aquaculturalists, according to Holmes?
2. What environmental impacts does aquaculture have, according to the author?
3. What comparison does the author make between America's prairies of the past and its coastal waters of today?

Reprinted from Bob Holmes, "The 'Blue Revolution,'" *New Scientist*, December 7, 1996, by permission of the *New Scientist*.

No one would mistake John Tucker's laboratory for the spray-soaked deck of a heaving North Sea trawler. But in a modest way he is beginning to take the place of a deep-sea fishing operation. In four Fiberglas tanks, thousands of tiny hatchling snook swim peacefully in filtered, sterilized water. In a few weeks, Tucker will transfer them to a much larger concrete tank, where in another year's time they will be big enough to catch and serve for dinner.

Based at Harbor Branch Oceanographic Institution in Fort Pierce, Florida, Tucker's operation gives a peek into what some say will be the future of fishing. The first agricultural revolution brought domestication of food crops and animals. Now another revolution is setting its hungry sights on the oceans. Already, one of every five fish destined for dinners worldwide comes from farms, and everyone expects that share to increase in coming decades. This "blue revolution," like the green revolution on land, promises greater efficiency and a steadier supply of fish.

But just as the green revolution has its dark side, so too has the revolution in aquaculture. The blue future may offer more water pollution, disease risks, loss of genetic diversity, and shortages of fresh water and other scarce resources. And not everyone will be able to afford the products of a new, improved, high-tech aquaculture.

THE CRISIS

Still, something must be done about the worldwide crisis in ocean fisheries today. After 40 years of steady increases, the world's wild fish catch has remained relatively constant at about 90 million tons since 1989. Stocks of cod, halibut, and other prime fish are dwindling; in 1996, several species made the World Conservation Union's Red List of species vulnerable to extinction. Canada's once bountiful Grand Banks cod fishery has been closed completely since 1992. Indeed, 25 percent of the world's wild fish stocks are now overexploited or have already crashed, according to the United Nations' Food and Agriculture Organization (FAO). Even the most optimistic experts think the wild-caught fish harvest can do no better than its present level over the coming decades. Pessimists foresee much worse. "You and I are probably members of the last generation who will sit down at dinner tables to things as exotic as grouper—or cod, even," says Jeffrey Graham, a fish biologist at Scripps Institution of Oceanography in La Jolla, California.

Everyone agrees, though, that in the future more and more of the fish we eat will come from fish farms. But if so, what will the future taste like? "I think instead of having all these different kinds of wild fish, they will be replaced by those fish that are most easily farmable," says Carl Safina, the New York–based director of the Living Oceans Program of the National Audubon Society. "Get ready to eat a lot of tilapia."

Tilapia—the chicken of the fisheries world—crop up often in conversations about aquaculture. Native to the lakes of Africa, these moderate-sized tropical fish grow quickly, adapt to fresh water or brackish lagoons, and tolerate incredible crowding. They will eat almost anything, from high-protein pellets to pond scum and even sewage. Behind the scenes, scientists are busily trying to breed a better tilapia by applying the same selective breeding skills that have worked so well with rice, wheat, and cattle. "The results are astonishing," says Meryl Williams, whose International Center for Living Aquatic Resource Management in the Philippines is a leader in the effort. After just four generations, the improved fish grow 60 percent faster and have a 50-percent higher survival rate to adulthood than the original tilapia.

But even low-tech fish farms often rear premium-price species. Many of the international agencies that lend money for aquaculture projects in developing countries insist that the farms raise high-value species such as shrimp or salmon, which can be exported to earn hard cash to repay loans. This bias means that cheaper fish that could be sold to local markets often never make it into large-scale culture.

There is another reason why the fish beloved of Western palates will never feed a hungry world, no matter how many we manage to farm, or how cheaply. In the wild, most of these species are predators near the top of the oceanic food chain, and they demand the same high-protein diet in fish farms. Today, that usually means fish meal—ground-up "trash fish" from the coast of Peru, Alaska, and other rich fishing zones. The FAO reckons that it takes a bit more than two pounds of fish meal to produce each pound of farmed ocean fish or shrimp. In other words, one of the enduring myths of aquaculture—that farming cod or salmon will "take the pressure off the oceans"—is simply false. It shifts the burden, or even adds to it. To conquer this problem, researchers are now trying to develop high-protein feeds that use soybeans or other non-animal sources of protein.

AQUACULTURE'S DRAWBACKS

However, intensive aquaculture on a large scale has its own drawbacks, just as large-scale agriculture does on land. When fish farmers flood artificial ponds, aquaculture can slurp up enough fresh water to create severe shortages. Thailand produced 250,000 tons of shrimp in 1994, but in parts of the country, fresh water used for shrimp ponds lowered local water tables by 13 feet in just two years.

Fish farms can also be serious sources of water pollution. A densely stocked fish cage is the equivalent of a cattle feedlot but set directly in a waterway, constantly pouring out uneaten food, feces, and other jetsam. In Thailand alone, some 197,000 acres of shrimp farms discharge about 1.7 billion cubic yards of effluent each year, according to a preliminary draft of a report on shrimp farming prepared for the FAO. A shrimp pond drained for cleaning, says one aquaculture expert, produced "the nastiest smell I've ever smelled."

Another problem aquaculture shares with agriculture is dealing with disease. Whenever large numbers of a single species are packed together—whether it's humans in cities, maize in fields, or shrimp in ponds—the risk of epidemics increases. In Taiwan, epidemics devastated the shrimp industry, which plummeted from a high of more than 90,000 tons in 1987 to almost nothing four years later.

Antibiotics can relieve bacterial diseases in the short term, but over time, the bacteria may simply become resistant. Geneticists have had some success at breeding strains of shrimp that are resistant to a few of the viruses that attack culture ponds, says Williams.

Even if aquaculture can cope with its problems of disease and pollution, a bigger issue remains, with moral as well as technical implications: If fish farms are to be the future of fishing, where will we put them all? About half the area now used for shrimp ponds in Thailand's Inner Gulf was once rice paddy. And China has grown concerned enough about loss of arable land that it now prohibits its conversion to aquaculture ponds.

A Trade-Off

Turning over the prairies to agriculture left one of North America's great ecosystems in tatters. Yet the cornfields that replaced them are now one of the world's most productive breadbaskets. Can we justify a similar trade-off in our coastal waters? And can we learn from experience and attempt a more balanced development? The world needs to decide soon, before unplanned exploitation races ahead.

PERIODICAL BIBLIOGRAPHY

The following articles have been selected to supplement the diverse views presented in this chapter. Addresses are provided for periodicals not indexed in the *Readers' Guide to Periodical Literature*, the *Alternative Press Index*, the *Social Sciences Index*, or the *Index to Legal Periodicals and Books*.

Erin Anderssen   "The Cod Collapse," UNESCO Courier, July/August 1998.

Peter Benchley   "Swimming with the Sharks," Audubon, May/June 1998.

Joel Bleifuss   "Save the Oceans—and the Fishermen," In These Times, July 22, 1996.

Stephen Budiansky   "Fish Farming Is Still on the Hook," U.S. News & World Report, August 5, 1996.

John Flicker   "A Grassroots Victory for Ocean Fishes," Audubon, March/April 1997.

Rodney M. Fujita, D. Douglas Hopkins, and W.R. Zach Willey   "Creating Incentives to Curb Overfishing," Forum for Applied Research and Public Policy, Summer 1996.

Anne Platt McGinn   "The (Aqua)cultural Revolution," World Watch, March/April 1998.

Steven Nadis   "Fertilizing the Sea," Scientific American, April 1998.

J. Madeleine Nash   "The Fish Crisis," Time, August 11, 1997.

Dennis W. Nixon   "Easing Pressures on U.S. Fisheries," Forum for Applied Research and Public Policy, Summer 1996.

Douglas S. Noonan   "Saving Commercial Fisheries," Different Drummer, Spring 1997. Available from the Thoreau Institute, 14417 SE Laurie, Oak Grove, OR 97267.

Carl Safina   "The Audubon Guide to Seafood," Audubon, May/June 1998.

Carl Safina   "Fisheries Management Improving," Issues in Science and Technology, Spring 1997.

Dick Victory   "Let Them Eat Fish," Washingtonian, April 1997.

Michael Weber   "The Fish Harvesters," E Magazine, November/December 1996.

CHAPTER 4

HOW CAN WHALES
BEST BE PROTECTED?

CHAPTER PREFACE

In 1946 the world's major whaling nations formed the International Whaling Commission (IWC) to regulate the whaling industry. Because of technological advances including sonar, harpoon guns, and factory ships, more whales were killed in the first four decades of the twentieth century than in the previous four centuries combined. The extensive hunting of whales was jeopardizing the survival of some whale species and the whaling industry itself.

To ensure the future of the industry, the IWC banned the hunting of certain species and set limits on the numbers of whales that could be caught. Its initial effectiveness was hampered by a lack of enforcement powers and by problems in ascertaining whale populations and setting realistic quotas. Meanwhile, international pressure to stop whaling altogether grew. Animal rights and environmental activists aggressively appealed to public opinion in seeking a worldwide ban on commercial whaling. Several countries, including the United States, passed laws outlawing whaling and the importation of whale products. Finally, in 1983, the IWC voted to enact a temporary ban on all commercial whaling beginning in 1986 (aboriginal groups who traditionally hunted whales for subsistence were allowed to continue). The moratorium was extended in 1991.

Some nations, notably Japan and Norway, have maintained whaling operations and have expressed strong objections to the IWC ban. Japan hunts several hundred whales annually for allowed scientific research purposes (meat is then sold on the Japanese market). Norway has authorized increasingly large harvests of minke whales, arguing that these relatively small whales are not endangered. Representatives of these nations argue that while some limits on whaling are necessary, whales can be hunted in a sustainable fashion. Whaling is a valued industry in some countries, many contend, and should not be banned entirely.

Supporters of the whaling ban argue that these animals remain in jeopardy because of ocean pollution, lack of food due to human fishing, and even increased ultraviolet radiation due to the thinning of the ozone layer in the earth's atmosphere. Many also criticize whaling as a harsh and barbaric practice against what they believe to be social and intelligent creatures. The viewpoints in this chapter debate the IWC whaling ban and other issues surrounding whales.

"If these magnificent denizens of the deep are going to remain, . . . immediate, decisive action is needed."

WHALES ARE GRAVELY ENDANGERED

Elaine Robbins

In the following viewpoint, freelance writer Elaine Robbins contends that although a hunting ban has helped stabilize world whale populations somewhat, whales are still threatened with a variety of human-made problems. Whales, she writes, are endangered by ocean pollution, underwater noise, collisions with ships, and the depletion of food sources caused by human fishing. Robbins argues that the global ban on commercial whaling should not be relaxed or lifted because of the continued precarious status of whales in the wild.

As you read, consider the following questions:

1. Which species of whales does Robbins believe are most endangered?
2. What potential effects does global warming have on whales, according to the author?
3. What steps does Robbins argue should be taken to save the whales?

Abridged from Elaine Robbins, "Whale Watch," E/The Environmental Magazine, May/June 1997. Reprinted with permission from E/The Environmental Magazine; Subscription Department: PO Box 2047, Marion, OH 43306; telephone: (815) 734-1242. Subscriptions are $20 per year.

In the early 1970s, the slogan "Save the Whales" became the mantra of the still-embryonic environmental movement. The plight of these beloved mascots of the sea captured the popular imagination—and mobilized protests that culminated in one of the great victories of green activism: a global moratorium on whaling. Now, 25 years later in 1997, "Save the Whales" is such a cliche—so '60s, really—that it's almost embarrassing to still be talking about it. But are the whales any better off now in the 1990s?

Unfortunately, the evidence suggests that they are not. Whales still haven't rebounded from the dramatic stock declines caused by centuries of hunting. Seven of the eight great whale species decimated by whaling are still on the Endangered Species List. Only one—the California gray, also known as the Pacific gray—has recovered in sufficient numbers to be removed from the list. . . .

WARNING SIGNS

But while activists' efforts have focused on holding the [whaling] moratorium, are we missing the warning signs that there are other, more immediate threats that could be sending some of the most endangered species toward extinction? Is a radically different approach necessary to truly save the whales?

Look at just a few of the signs:

• The Northern right whale—so named because it was slow and yielded plenty of oil and was therefore the "right" whale to hunt in the early days of whaling—is the most endangered species. Only 295 remain; so few that biologists have named each of them. The greatest current threats to its existence are not hunting but human encounters. The right whale faces collisions with cargo ships and entanglement in fishing nets as it commutes up and down the eastern U.S. coast between New England and northern Florida.

• The St. Lawrence River population of beluga whale is so contaminated by DDT and PCBs that when one dies its carcass has to be disposed of as toxic waste. This population, which once numbered 5,000, has been reduced to 500.

• The Southern Hemisphere population of the blue whale, the largest mammal on Earth, could become extinct within the next decade. That's no small thing, even in an age when species drop off the face of the planet with alarming regularity. There are so few left—just a few hundred animals, down from an original 250,000—and they travel over vast stretches of ocean, that they may be having trouble finding mates. Marine biologists fear that the growing cacophony of manmade sounds under the sea—everything from ship engines to oil drills—may be interfering with their ability to communicate with each other over long distances.

It's hard not to notice that whales have been washing ashore sick and disoriented with increasing frequency in the last 20 years. The likely culprit? Water pollution. "Sooner or later, all persistent chemicals released into the environment end up in the oceans," says World Wide Fund for Nature Senior Scientist Theo Colborn. Whales tend to accumulate these toxins in their blubber by eating fish tainted with mercury, lead, and PCBs—substances that many scientists believe have caused everything from immune system and metabolic problems to decreased fertility and genetic mutations in other species.

Although scientists can't predict what effects this exposure will have in the long run, they fear that the effects could be drastic and leave little time for recovery. "Whale populations could crash suddenly with little warning," says Colborn.

THE CRISIS IN KRILL

With man's insatiable taste for fish and seafood, whales now find themselves competing with humans for food. With the depletion of 13 of 17 of the world's major fish stocks, fish-eating species are having to spend more time hunting for food. Baleen whales—humpbacks, blues, fin and minkes—have until recent years been saved by their penchant for eating low on the food chain: krill is their main dietary staple. But now humans are developing a taste for this tiny shrimplike creature, too. Krill is increasingly being harvested as a "natural supplement" and is sold frozen, canned and as a food paste in supermarkets throughout Asia and the Pacific. . . .

Whales may also be among the first animals to feel the effects of ozone depletion and global climate change. Cases of skin cancer have been turning up in species that live in Antarctic waters beneath the hole in the ozone layer. There is also evidence that increased UV exposure could hurt Antarctic krill, which live in the upper part of the water column, making them vulnerable to UV exposure. Krill eggs, which float on the surface, are even more vulnerable. Scientists at the Southwest Fisheries Center have found that UV exposure can damage the eggs of small fish—an effect it's likely to have on krill eggs as well.

And if Antarctic ice continues to melt as a result of global warming, whale habitats could change dramatically. Says Michael Tillman, director of the Southwest Fisheries Center, "Antarctic blue whales feed on concentrations of krill that live along the ice edges and in the ice pack. If their habitat is changed—loss of ice, loss of food—then they have nowhere else to go. And that's very worrisome."

But the biggest problem is simply that low numbers make recovery a slow process. "We're seeing very little progress so far in a relatively short time span," says Gerald Leape, Greenpeace's legislative director for ocean issues. "Fish can bounce back in a couple of years. Whales take longer than that. Whales reproduce more slowly than humans."

SHOULD WHALING BE PERMITTED?

Given these new threats, does it make any sense to reopen the seas to whaling? Is there any valid reason to refuse Norway's request to the International Whaling Commission for permission to conduct a "sustainable" hunt of abundant minke whales? (Whalers went after the largest whales first, so the minke, the smallest of the great whales, was spared the devastation of the other species.)

There's only one problem: Whalers have a history of cheating. In 1993 the Soviet Union stunned the world with the revelation that, throughout the 1960s and 1970s, they had killed perhaps as many as 20 times the number of the most endangered species than they had previously reported. The blue whale in particular offers an irresistible lure for whalers who might be tempted to cheat. At 60 to 80 tons and more than 100 feet long, a blue whale could fetch as much as half a million dollars on the Japanese market.

To understand the problem, visit any fish market in Japan, where you'll find a bewildering variety of fish and seafood for sale, from bluefin tuna as big as a man's torso to buckets full of live eels. But most mysterious is the whale meat that's sold for up to $160 a pound and usually ends up as sashimi in expensive Japanese restaurants. Selling some whale meat domestically is legal—the Japanese continue to hunt a few hundred minke whales each year under a "scientific purposes" loophole in the moratorium and are permitted to sell the meat at home. But you can never be too sure what you're looking at. In 1993, two biologists from New Zealand and Hawaii performed DNA tests on a variety of whale meat purchased at retail markets throughout Japan. What did they find? Many of the samples were not minke but such endangered species as humpbacks, fin whales and blue whales, and some was illegally imported.

In fact, some environmentalists fear Norway's legal hunt provides a cover for the illegal trade. Contraband shipments of whale meat are seized periodically entering Japan and Korea. Whaling nations could get such an opening as they mount the biggest attack on whaling legislation since the onset of the moratorium at the Convention on International Trade and Endangered Species (CITES) meeting in Zimbabwe June 9–20, 1997. Japan and Norway have filed resolutions to reclassify six different species of great whales from Appendix I to Appendix II, from threatened status in which trade is strictly prohibited to one in which trade is regulated but allowed. [The resolutions did not receive enough votes to pass.]

"Save the blue whale . . . save the blue whale . . . save the blue whale . . ."

© Punch/Rothco. Reprinted by permission.

Says Greenpeace's Leape, "If they're successful, it will give them a legal justification for pushing a resumption of commercial trade. And if they get a resumption of commercial trade, then they lift the moratorium, and then commercial whaling begins again—not just by Norway, but on a large scale. That will allow whalers to reopen the markets with Japan, and then there's that economic incentive. Because right now, there's no other country in the world that offers that kind of price for whale meat and whale products."

"We've stopped the decline, but the whales need some time to recover," says Leape. "And the last thing we need is a full-scale resumption of commercial whaling—that will really drive them down the tubes."

WHAT SHOULD BE DONE?

So what should we do now if we truly want to save the whale? Many conservationists celebrated the 1994 creation of an Antarctic whale sanctuary along the whole bottom third of the Earth as a great victory. But sanctuaries may not be the answer. (As Jacques Cousteau said in his inimitable way, "Our God! Was it forgotten that whales migrate?") But we do need to continue to hold the line against commercial whaling, especially against the kind of critical assault that's being mounted in Zimbabwe. Cousteau has suggested extending the moratorium another 50 years—the period he says is needed for populations to truly rebound to allow a "maximum sustainable yield." In 1996 the Australian government announced its plans to push for a permanent ban on commercial whale-hunting.

Much can be done to improve monitoring and enforcement of the whaling that does continue. We need to improve the reliability of the species counts upon which whaling limits are based. We need a more effective observation and inspection system. (The World Wildlife Fund has proposed that each whaling vessel at sea longer than 12 hours have an inspector on board.) DNA testing could be implemented on a widespread basis to enforce whaling laws. . . .

To save the critically endangered Northern right whale, the U.S. government needs to take the lead in introducing measures to protect this slow-moving creature from ship collisions. "These whales are predictable," says Leape. "You can tell they're going to be off Charleston during the month of March and off New England in May, June and July. So you could say that during specific times of year—say if you're a ship that has to go into Charleston in March—that you cut your speed by half and take other precautions.". . .

Most important, we need to start finding ways to protect the oceans, of which whales' decline is only the most visible symptom. Along with continued efforts to reduce water pollution, measures should be introduced into international maritime law that would set limits on undersea noise. To accommodate such limits, available technology could be used to make ship engines less noisy.

Cetaceans continue to awe and inspire the human species, as the ever-growing list of whale watches attests. But if these magnificent denizens of the deep are going to remain as fellow Earth travelers—and not just displays of dusty bones in the world's museums—immediate, decisive action is needed. "Save the Whales" will have to be a rallying cry again.

| "To say that 'whales' are endangered is no more meaningful than to say that 'birds' are endangered."

THE ENDANGERED STATUS OF WHALES HAS BEEN EXAGGERATED

David Andrew Price

In the following viewpoint, David Andrew Price asserts that supporters of a continued global ban on commercial whaling rely on deceptive and misinformed arguments. Foremost among these misconceptions is the belief that whales are an endangered species. While several species of whales remain threatened, many others are not, Price contends, including the minke whales that Norwegian and Japanese whalers are seeking permission to hunt. A blanket universal ban on whaling has no basis in science or conservation, he concludes, but simply reflects the cultural prejudices of whaling opponents. Price is a writer and attorney with the Washington Legal Foundation.

As you read, consider the following questions:
1. What improper activities of anti-whaling activists does Price describe?
2. What is the world population of minke whales, according to the author?
3. What motivates American policy on whaling, according to Price?

Abridged from David Andrew Price, "Save the Whalers," *The American Spectator*, February 1995. Reprinted with permission. © The American Spectator.

One morning last January 1994, Arvid Enghaugen, a resident of the Norwegian coastal town of Gressvik, found his whaling boat sitting unusually deep in the water. When he climbed aboard to investigate, he found that the ship was in fact sinking; someone had opened its sea cock and padlocked the engine-room door. After breaking the lock, Enghaugen discovered that the engine was underwater. He also found a calling card from the Sea Shepherd Conservation Society, a small, California-based environmentalist group that specializes in direct actions against whalers. Counting Enghaugen's boat, Sea Shepherd has sunk or damaged eleven Norwegian, Icelandic, Spanish, and Portuguese vessels since 1979.

The boat was repaired in time for the 1994 whaling season, but Enghaugen's problems weren't over. On July 1, while he was looking for whales off the Danish coast, five Greenpeace protesters boarded the ship from an inflatable dinghy and tried to take its harpoon cannon. Enghaugen's crew tossed one protester into the sea, and the rest then jumped overboard; the protesters were picked up by the dinghy and returned to the Greenpeace mother ship. . . .

Although the activists failed to stop Enghaugen's hunt, their public relations war in America has been a different story. Over the past twenty years, the save-the-whales movement has been so successful in shaping public sentiment about the whaling industry that the U.S. and other nations have adopted a worldwide moratorium on whaling. Part of the credit must go to the animals themselves, which are more charismatic on television than Kurds, Bosnians, or Rwandans, who have engendered far less international protection. The movement owes most of its success, however, to the gullibility of Hollywood and the press in passing along bogus claims from whaling's opponents.

MISINFORMATION

The mainstay of the case against whaling—that it threatens an endangered species—is characteristic of the misinformation. It is true that European nations and the United States killed enormous numbers of whales during commercial whaling's heyday in the nineteenth century, but to say that "whales" are endangered is no more meaningful than to say that "birds" are endangered; there are more than seventy species of whales, and their numbers vary dramatically. Some are endangered, some are not. The blue whale, the gray whale, and the humpback were indeed depleted, but those species were later protected by international agreement long before the existence of Greenpeace or Sea Shepherd. (There have been abuses. Alexei V. Yablokov, special adviser to the president of Russia for ecology and health, has revealed that the whaling fleet of the former Soviet Union illegally killed more than 700 protected right whales during the 1960s, but the International Whaling Commission's institution of an observer program in 1972 essentially put an end to the Soviet fleet's illegal activities.)

SCIENTIFIC REALITIES

The International Whaling Commission has always been slow to react to political and scientific realities. . . .

In the late 1960s, whale populations around the world were suffering from over a century of nearly unfettered exploitation, and critics quite rightly pointed out that the IWC had failed to stem this decline. When environmentalists finally wrested control of the IWC in 1982, the result was a complete ban on commercial whaling that went into effect in 1986. Today many of the world's great whale species are thriving, but the IWC remains in denial. . . .

The case of the diminutive minke (the smallest of the baleen whales at only 30 feet long) is a good example. Some years ago the IWC scientific committee estimated the worldwide minke whale population at about one million animals, even more abundant than before commercial whaling began. Remarkably, the IWC still considers them threatened with extinction.

Michael De Alessi, *Asian Wall Street Journal*, October 17–18, 1997.

The only whale species that Enghaugen and his fellow Norwegian whalers hunt is the minke, which Norwegians eat as whale steaks, whale meatballs, and whaleburgers. As it turns out, minke whales are no more in danger of extinction than Angus cattle. In 1994, thirty-two Norwegian boats killed a total of 279 minkes, out of an estimated local population of about 87,000 and a world population of around 900,000.

In 1982 the IWC voted to suspend commercial whaling for a five-year period starting in 1986. The ostensible purpose was to permit the collection of better data on whales before hunting resumed. Norway lodged a reservation exempting itself from the moratorium, as the IWC treaty permitted, but it complied voluntarily.

AN INDEFINITE SUSPENSION

Whaling nations soon learned, though, that the majority of nations in the IWC—including the United States—intended to maintain the ban indefinitely, no matter what the numbers showed. Canada left the IWC in 1982, and Iceland left in 1992. Norway terminated its voluntary compliance in 1993. To protest the commission's disregard of the facts about whale stocks, the British chairman of the IWC's scientific committee resigned that year, pointing out in his angry letter of resignation that the commission's actions "were nothing to do with science." The IWC continued the moratorium anyway at its next meeting.

A 1993 report by the Congressional Research Service observed that the data on whales undercut the conservationist argument, and that "if the United States argues for continuing the moratorium on commercial whaling, it may have to rely increasingly on moral and ethical appeals." The ban on whaling is no longer about conservation, in other words, but about the desire of many Americans and Western Europeans to impose their feelings about whales upon the whaling nations (which include Iceland, Russia, Japan, and the Inuits of Canada and Alaska).

ARE WHALES INTELLIGENT?

Popular notions of whales' human-like intelligence, often cited by opponents of whaling, have little real support. Whales possess large brains, but that proves nothing about their mental agility. Margaret Klinowska, a Cambridge University expert on cetacean intelligence, holds that the structure of the whale brain has more in common with that of comparatively primitive mammals such as hedgehogs and bats than with the brains of primates.

Whales can be trained to perform stunts and other tasks, but so can pigeons and many other animals that have never been credited with the cerebral powers of Homo sapiens. And the idea that whales have something like a human language is, at present, pure folklore. Like virtually all animals, whales make vocalizations, but there is no evidence that they are uttering Whalish words and sentences. Their famed "singing" is done only by the males, and then during but half the year—a pattern more suggestive of bird-song than human speech. . . .

What American policy on whaling enforces is simply a cultural preference—one comparable to our distaste for horsemeat, which is favored in France. The whale-savers have succeeded in shaping policy by selling the idea that whales are different: that they are endangered underwater Einsteins. That's why Icelandic filmmaker Magnus Gudmundsson, who has produced a documentary showing Greenpeace's machinations on the issue, is correct in calling the movement "a massive industry of deception."

"The history of modern commercial whaling is one of repeated over-exploitation."

THE BAN ON WHALE HUNTING SHOULD CONTINUE

Greenpeace

Commercial whaling was banned by the International Whaling Commission in 1986 because continued hunting was believed by many to cause the extinction of several whale species. Greenpeace, an international environmental organization, has long opposed commercial whaling. In the following viewpoint, the organization argues that the ban should remain in effect because resumed whale hunting would threaten slowly rebuilding whale populations already endangered by past hunting practices and by environmental pollution.

As you read, consider the following questions:

1. What main differences exist between whales and fish, according to the organization?
2. Why does Greenpeace consider whale counts to be unreliable?
3. Why is sustainable whaling not possible, according to Greenpeace?

Reprinted from "The Case Against Whaling," an undated paper on the Greenpeace website, www.greenpeace.org/~comms/cbio/case.html, with permission from Greenpeace.

The history of modern commercial whaling is one of repeated over-exploitation, as whaling companies strove to maximize their short-term profits despite the longer-term implications for the very species on which their livelihood depended.

Now, past mistakes are being acknowledged by the modern-day whaling lobby in their campaign to convince the world that whaling today will be different and will be done in a sustainable manner for the first time. In fact nothing has changed.

The relatively few whales that remain are highly vulnerable, and the factors that led to their over-exploitation in the past have not changed. Here are the key arguments against whaling.

Whales are not fish.

Whales are mammals, not fish, but they have historically been treated as fish by the whaling industry.

The vast majority of fish species reproduce by releasing huge quantities of eggs into the water for fertilization by the male, although under normal conditions only a small percentage of these will develop into mature adults. Whales on the other hand have a long gestation period and usually give birth only every one or two years, to a single calf that requires more than a year of maternal care before it can survive on its own. Even then, whale calves take many years to reach maturity. For these reasons whales can never recover quickly from exploitation.

These factors are compounded by our lack of knowledge about many aspects of whale biology. Even after decades of research, the growth rate of whale populations is unknown because of the difficulty of studying these highly migratory, long-lived, slow-reproducing animals. Nor are there reliable estimates of live birth rates or of the natural mortality rates of calves and juveniles.

Fishing industry representatives who stir up concern that a ban on whaling would lead to a ban on fishing are deliberately confusing the public and disregarding the vastly different characteristics of two entirely unrelated groups of species.

COUNTING THE WORLD'S WHALES

We can't count whales accurately.

Modern scientific methods cannot count whales accurately. Determining how many animals exist in a population is crucial for any calculation of kill quotas. The size of most populations of whales is known no more accurately than plus or minus 50 percent. Since populations change so slowly, it is impossible to tell if a population is growing or shrinking in the course of a few years' study.

All population estimates are based on a count of the whales sighted on each side of a survey vessel as it zigzags its way through a designated stretch of water. Since only a small percentage of the whales in any given population will be visible on the surface as the vessel passes, extrapolations must be made from the number sighted to give an estimate for the entire region under study. Thus all population estimates are based on sightings of a tiny fraction of the population.

In 1995 a Norwegian minke whale survey sighted 29 minke whales in the eastern Barents Sea. Based on this they calculated a population estimate of 16,101—over 500 times greater than the number of whales seen.

Mathematical formulas are used to calculate the total number of whales from the small number of actual sightings. These formulas attempt to take into account numerous variables. It is in these formulas that an enormous potential for compounded error exists.

For example, in the method used by Norway the final population estimate is very sensitive to the accuracy of the observers' estimate of how far the whale was from the ship when it was seen. If the distance estimated is half the true distance, then the population estimate will be four times the true population. So if observers thought the whales which were 800 meters from the ship were 400 meters away, the calculations would estimate a population of 30,000, not 120,000.

The weakness of these estimates is well documented. In 1986 a joint Japanese/Soviet survey of one area of the Antarctic estimated 56,000 minke whales; a Japanese survey of the same area 5 years later estimated just 2,800. This does not mean that the whales had declined; it simply illustrates how uncertain these numbers are.

Whale counts are not censuses. They are estimates based on extrapolation and highly uncertain.

THE ECONOMICS OF WHALING

Sustainable whaling is economically unsound.

Common sense would seem to suggest that because whale populations grow very slowly, it would be in the long-term interests of the whaling industry to maintain whale populations at a healthy level, and to avoid over-exploitation that might lead to a terminal decline.

But mathematician Colin Clark showed, in a classic paper called "The Economics of Overexploitation" published in the journal *Science* in the early 1970s, that exploitation of slow-growing populations will naturally lead to their severe depletion and even to their extinction. This is because if the profit to be had from catching the entire population of animals at once is greater than the profit that could be made by conserving the population and taking only a certain quantity each year, then it is good business to catch as many animals as possible, as quickly as possible—even though it is bad management from the biological point of view.

Economic logic, divorced from the realities of the living world, dictates that high levels of short-term exploitation will bring the best financial return. In other words, sustainable whaling is economically unsound.

© Raeside/Rothco. Reprinted by permission.

Killing more whales won't restore the ecological balance.

There is little doubt that overfishing has devastated fish stocks in many parts of the world. The whaling lobby is now arguing that this means that whales should be killed to protect the remaining fish.

In fact, whales are far from being the only consumers of fish; huge quantities are eaten by other fish and by seabirds. In any case, killing a whale does not release the fish that it would have eaten to a commercially valuable fish. It is just as likely to be eaten by another species altogether. There is not a single case worldwide where it has been demonstrated that a catch of whales has increased the take of commercially valuable fish.

WHALES AND THEIR ENVIRONMENT

Whales live in a changed ocean.

Whales evolved over tens of millions of years and are superbly adapted to the marine environment. But this environment is now under a sustained threat from the consequences of human activities, including climate change, increased ultraviolet-B (UV-B) radiation from a diminishing ozone layer, pollution by organochlorine compounds and increasingly intensive fisheries.

Organochlorines have been shown to cause reproductive failure in marine mammals and to be potent suppressors of mammals' immune systems. Whales and other marine mammals concentrate these artificial compounds through the food chain so concentrations in the animals may be a million times higher than in the surrounding seawater. Recent work suggests that some classes of compounds, called endocrine disrupters, exert very powerful negative effects on reproduction at extremely low concentrations.

UV-B can affect the plankton which are at the base of the ocean food chain. Some species are extremely sensitive and others are more resistant, so species composition may change. In some species UV-B damages the sensory cells that respond to light and gravity, causing them to swim randomly and lose their optimum position in the water column.

Even the most oceanic of whales no longer live in a pristine environment. Anything less than a precautionary approach is unacceptable in the effort to conserve whales. The benefit of the doubt must be given to the environment. A failure to do so may extinguish any hope of long-term protection for the world's whale populations.

"Officials . . . need to remember that
whales and whalers can
coexist—and indeed they must."

THE BAN ON WHALE HUNTING IS
NOT JUSTIFIED

William Aron

William Aron is a former U.S. delegate to the International
Whaling Convention (IWC) and member of the IWC's Scientific
Committee. In the following viewpoint, he argues that the commercial whaling moratorium imposed by the IWC in 1983
(which took full effect in 1986) is an extreme policy that is not
warranted by a rational consideration of the facts. According to
Aron, many whale species are not in danger of extinction and
therefore should be open to hunting. Furthermore, Aron asserts
that the moratorium unfairly penalizes the people who make
their living from the sustainable hunting of whales.

As you read, consider the following questions:

1. What role did the United States play in instigating the ban on
   commercial whaling, according to Aron?
2. In the author's opinion, why do American politicians support
   the whaling ban?
3. What important aspect of environmental management does
   Aron say whaling activists overlook?

The campaign to "save the whales," perhaps the most successful animal-protection movement ever, had its origin in a problem described in Garrett Hardin's classic 1968 essay "The Tragedy of the Commons": Like many natural resources, whales aren't owned by anyone but are held in common by all. In such situations, resources can be exploited by a few individuals, harming all other current and future users. This can lead to truly tragic consequences, and indeed most natural-resource managers are aware, through experience, of the accuracy of Mr. Hardin's analysis.

We are even more painfully aware, however, of the problems associated with attempting to avoid such tragedies. This is especially true with whales. Relatively few people have any direct involvement with these marine mammals, whether directly in the whaling industry or indirectly as fishermen competing for fish in the same waters. These professionals, however, must contend with policies determined by the rest of the world—which knows little about whales, but which can be awfully emotional about these majestic animals.

The U.S. delegation to the International Whaling Commission (IWC) . . . has a chance to help right some of the wrongs of international whaling regulation. Perhaps most importantly, it can defend Japan's and Norway's annual harvest of several hundred minke whales each against claims by Sea Shepherd, Greenpeace, the International Fund for Animal Welfare and other environmental extremists who would end whaling altogether. And the U.S. can also press the claims of the Makah Indian tribe to resume traditional whale hunts off the coast of Washington state.

The U.S. commissioner to the IWC, D. James Baker, has a responsibility to act decisively, for it was the U.S. that took the lead in calling for a moratorium on whaling at the 1972 U.N. Conference on the Human Environment in Stockholm. The IWC, which began in 1947 as a whalers' cartel, at first refused to implement the moratorium. But environmentalists and animal-rights activists exerted immense pressure on the IWC and its member nations, and by 1983 the IWC had adopted a "moratorium" on commercial whaling: In order to hunt whales, member countries must receive specific exemptions from the commission.

This extreme policy was supported by the U.S., whose politicians had discovered that by supporting whale protection efforts, they could demonstrate their environmentalist credentials without losing the support of labor unions, property developers or other constituents. Animal-protection activists had persuaded the public that all whales are endangered, that whales are uniquely intelligent, and that more attractive alternatives are readily available to those businesses and communities that derive their livelihoods from whaling.

The problem is that none of these widely held beliefs is true.

First, it is true that some whale species are endangered. But besides the bowhead whale, which is harvested by indigenous peoples of the U.S., Canada and perhaps Russia, and some fin whales that may be taken by the indigenous people of Greenland, no IWC members are proposing to harvest any endangered whales.

The minke whale, which has never been endangered, has a world population of more than one million, according to the IWC's Scientific Committee. It is this species that whalers of Japan, Greenland and Norway are harvesting. The gray whale, which was removed from the endangered species list in 1994, has a population of more than 20,000. The sperm whale is on the U.S. endangered species list, but its exploitable worldwide population (all males older than 13 years and females older than 10) is more than one million, according to the IWC and the National Marine Fisheries Service. The total sperm whale population is probably twice that number. Endangered? Hardly.

WHALES ARE SACRED COWS

A lot of my friends are going to disapprove, but I think it's time to legalise the killing of whales again. They have become the sacred cows of environmentalism. . . .

Environmentalists who insist the ban should stay are blindly sticking to an out-of-date position, in defiance of scientific evidence.

Geoffrey Lean, *Independent on Sunday*, October 19, 1997.

Second, whales are unquestionably extraordinary animals. While they possess a certain intelligence, however, there are no data to support the belief that they are at or even near the top of the animal intelligence scale. Bees, wasps, ants, birds, mammals and many other animal groups also demonstrate complex social behavior, communicate with one another and exhibit learning skills, yet no one seriously proposes that we protect them from any and all human activities.

Third, the people in Japan, Greenland, Norway and Iceland who make their livings from whaling have no ready, cost-effective alternatives. Given the abundant worldwide whale populations, there is no reason to penalize the many in these countries who choose to toil—skillfully and at high risk—to hunt a few thousand whales per year.

Closer to home, the Makah tribe is seeking to resume its harvest of gray whales, using traditional methods, now that the species is no longer endangered. The tribe pledges to take only five whales per year, and not to sell any of the whale meat or blubber. Yet many environmental activist groups are opposed even to this level of whaling—even though the IWC allows Russia's aboriginal peoples to hunt 169 gray whales each year.

There's another important aspect of environmental management that whale-protection activists overlook: Selective protection of one element of an ecosystem inevitably changes the entire system. In recent years evidence has appeared that even though the endangered blue whale has not been hunted for many decades, its populations haven't recovered in the Antarctic. Rather, it seems likely that seals, minke whales and other species have filled the biological niche that the blue whales once occupied. Some harvesting of the blue whales' competitors may well be needed if that endangered species is to recover. Similarly significant unintended consequences can follow any attempt to "protect" an ecosystem.

Officials from the State, Commerce and Interior Departments, the Marine Mammal Commission and other agencies . . . need to remember that whales and whalers can coexist—and indeed they must.

"Every *Alliance* member threads conservation themes throughout its park and education programs, acquainting people . . . with the importance of caring about marine mammals and the oceans that sustain them."

MARINE THEME PARKS THAT KEEP CAPTIVE WHALES PROMOTE CONSERVATION

Alliance of Marine Mammal Parks and Aquariums

Some animal rights and environmental organizations have campaigned against the captive display of marine mammals, especially dolphins and whales. In the following viewpoint, the Alliance of Marine Mammal Parks and Aquariums (AMMPA) defends this practice, arguing that marine parks and similar organizations contribute to public education, scientific research, animal rescue, and ocean conservation efforts. The alliance contends that by enabling the public to see and learn about whales and other creatures that most people would never see in the wild, marine parks help raise public awareness and support for marine wildlife conservation. The AMMPA represents approximately forty organizations that work with captive marine mammals.

As you read, consider the following questions:

1. What did a public opinion poll reveal about support for marine parks, according to the organization?
2. In the view of AMMPA, what advances in research have marine parks helped facilitate?

Reprinted from the Alliance of Marine Mammal Parks and Aquariums, "The Mission of Zoological Parks and Aquariums and How They Help Conserve Marine Mammals," available on www.pbs.org/wgbh/pages/frontline/shows/whales/debate/intro.html. Reprinted with permission.

The history and significance of the public display of marine mammals is a compelling story.

It has been just a short 25 years since experts realized that something had to be done to protect marine mammals and their habitats. At that time, whale and dolphin populations were being decimated around the world by commercial fishing, pollution, and human indifference to the consequences of their individual actions. As hard as it might be to understand today, people didn't care about these animals.

In the U.S., our Congress responded by passing the Marine Mammal Protection Act in 1972, which, among other actions, helped stimulate advances in fishing practices that have resulted in a huge reduction in the number of marine mammals incidentally killed in fishing nets. In 1972—in one year—an estimated 400,000 dolphins died as a result of the tuna purse seine fishery in the eastern tropical Pacific ocean alone. In 1995, that number dropped to approximately 3,000.

Our legislators also recognized that the public needed to learn about these animals and that greater knowledge was necessary to protect them. Congress urged public display facilities to foster public support for the conservation of marine mammals through education programs and to invest in research to increase our scientific understanding. Lawmakers understood that limiting the extraordinary contributions of the public display community to the conservation of these animals would have long-lasting and irreparable, negative results for the country's marine wildlife and the people of United States.

TOUCHING HEARTS AND MINDS

Along with community organization awards and citations, letters of praise from educators, visitors, parents, and young people attest to the impact Alliance education programs are making each day on our youth. Many document how this learning experience stimulated themselves, their children, or their students to become involved with the conservation of marine ·mammals or the environment for the first time. After returning home, some teach courses, others organize programs, present classroom projects, or write articles—all with a conservation theme. One mother's letter about her son's experience at a marine park crystallizes the legacy Alliance educators strive to provide the children they reach.

She wrote, "Whatever he does with his life, his interest in science was further enhanced and will be with him forever. I believe that when he is grown, I'll remember his trip as one of those crucial events, which put him on the right track to becoming an effective and responsible adult."

In the marine mammal community, we passionately care about marine mammals and share them every day with thousands of children around the world, touching their hearts and minds so they will learn to care as deeply as we do about the conservation of marine wildlife. We believe that each child we reach can make a contribution to a better environment for his or her children.

The people, in turn, believe in us. Marine life parks, aquariums, and zoos have strong and committed support from the public. A 1995 Roper poll shows that nine out of 10 people believe these institutions are essential to educating the public about marine mammals. Eighty-seven (87) percent say they wouldn't have a chance to see these animals if they could not visit a public display facility. The Canadian Advisory Committee on Marine Mammals commissioned a similar poll in 1992 from Decima Research. The results were equally positive.

The Alliance of Marine Mammal Parks and Aquariums was founded, in part, to share the latest advances in marine mammal care and husbandry and to foster responsible management practices. Member facilities have a common bond—a personal commitment to these animals, to assure that they have the very best quality of life.

The Alliance is an international association representing approximately 40 marine life parks, aquariums, zoos, scientific research facilities, and professional organizations. Member institutions are dedicated to the conservation of marine mammals and their environments through public display, education, and research. Although the Alliance headquarters are in the United States, almost 25 percent of our member facilities are located in other countries around the world. Through the Alliance, public display facilities are making a difference by getting this conservation message to the public as well as legislators, regulators and the media, opinion leaders that can make a difference for marine mammals and their environments.

Each Alliance member is dedicated to conservation, education, and research. Let's consider each of these areas for a moment.

CONSERVATION

Every Alliance member threads conservation themes throughout its park and education programs, acquainting people, especially children, with the importance of caring about marine mammals and the oceans that sustain them. The fundamental need for this information was confirmed in the Roper poll. The poll indicates that most people believe that the more they learn about these animals, the more likely they are to fight to preserve them for generations to come.

Alliance members offer over 1500 special lectures, courses, and programs with specific conservation themes for both adults and children each year. Topics range from coral reef ecology, oil spills, and ocean pollution to endangered species, marine debris, stranded animals, and specific ecosystems.

In conjunction with government agencies, Alliance members throughout the world work with manatees, vaquita [Gulf of California porpoises], river dolphins, stellar sea lions, California sea otters and Hawaiian monk seals—all species that are threatened or endangered—working to stave off their extinction.

Without marine life parks, aquarium, and zoos, to whom would governments turn to rehabilitate declining species like the Hawaiian monk seal pups that must be weaned and cared for, sometimes a year or more? Who would help assess the impact of environmental contaminants on immune and/or reproductive systems of wild populations, such as St. Lawrence belugas, as Alliance members have done for the Canadian government? Who would house and care for the sea lions that have been jeopardizing the steelhead salmon stocks in Washington state?

EDUCATION

Over 36 million adults and children walk through our institutions each year and are exposed to the education information provided to the general public through graphics, presentations, exhibits, and narrations about marine mammals and other wildlife. But, we don't stop there. About 100 million children, adults, and teachers are reached annually through specially designed, on-site education programs and educational messages distributed through computer learning programs, publications, teacher aids, satellite television, and outreach programs supported by our institutions.

Without marine life parks, aquariums, and zoos, who would effectively educate our children, and their children, to care about marine wildlife? Who would fund the learning materials, posters, booklets, fact sheets, and videos or offer teacher training, college courses, and curriculums for science courses? Who would garner the respect for these animals that engenders a strong, active commitment to marine mammal conservation, for which the public must ultimately shoulder the responsibility?

RESEARCH

In the area of research, it cannot be emphasized enough that much—if not most—of what we know about marine mammals today has been learned from research at public display facilities.

Researchers were never certain of a killer whale's length of pregnancy until reproductive studies could be conducted at marine life parks. Yet this information is vital to understanding the animals' ability to sustain a healthy population in the wild.

Today it is possible to perform a complete diagnostic ultrasound body scan on a wild dolphin in less than 15 minutes. This technique has been used at the request of our government in research studies to assess the health of populations troubled with mass strandings or disease. Without years of being able to study animals in aquariums and parks, without the ability to adapt human technology to animals in controlled environments, this extraordinary, noninvasive technique and others would simply not be available today to study wild marine mammal populations.

BREEDING PROGRAMS AT SEAWORLD

SeaWorld has learned much about the gestation and birth of killer whales through the parks' breeding program—the most successful of its kind in the world. As of December 1996, 11 calves had been born and are thriving at the four SeaWorld parks.

In 1993, zoological history was made at SeaWorld of Texas with the birth of the world's first second-generation killer whale calf born in a marine life park. The calf's mother, born at SeaWorld of Florida in 1985, was the first killer whale born in the care of man.

Much important biological and behavioral information is gathered from SeaWorld's killer whale breeding program. SeaWorld scientists have confirmed that the gestation period of killer whales, once thought to be 12 months, is actually 17 months. This information has allowed population biologists to more accurately estimate the dynamics of killer whale populations in the wild. SeaWorld routinely shares information about the development and physiology of killer whales with the public and the scientific community.

SeaWorld's Mission: A Commitment to Marine Life, 1996.

Alliance members have spent an estimated $20 million on research from 1992 to 1997—research that is essential to understanding the behavior, anatomy, and physiology of marine mammals; to rehabilitating stranded animals; aiding the conservation of wild populations; and to learning to better manage and assist endangered species.

Additionally, many Alliance facilities make their animals available to marine mammal researchers from colleges, universities, and other scientific institutions conducting noninvasive studies important to the animals' conservation and health. Much of this research simply cannot be accomplished in ocean conditions.

These studies have led to the development of vaccines and new methods of treatment; improvements of techniques for anesthesia and surgery: tests for toxic substances and their effects on wild marine mammals; and advancements in diet, vitamin supplements, and neonatal feeding formulas.

However, there is still a tremendous amount we do not yet know. And we desperately need greater knowledge and understanding if we are going to make informed, intelligent decisions regarding the increasingly complex pressures on our wild dolphins and whales. But we will never have a chance of gaining this knowledge and understanding without the opportunity to continue crucial studies at public display facilities.

Without public display facilities, who would cooperate with a university studying manatee energetics to determine why manatees are vulnerable to cold weather? This information is critical in determining when Alliance members can release hand-raised or orphaned animals safely into their natural habitat. Who would participate in government studies to prevent marine mammals from getting entangled in gillnets or fund augmenting work on noise makers, which could alert the animals to the presence of the nets?

RESCUE AND REHABILITATION

Another crucial role of Alliance members is giving a hand and hope to stranded animals. Over 1600 marine mammals were rescued, rehabilitated, and released from 1992 to 1997 by our member facilities, which voluntarily participate in stranding networks. The medical advances and techniques developed through our research is a huge benefit to these sick and injured animals as we struggle to save their lives and return them to their natural habitats.

Alliance members are not reimbursed for the dedicated care they provide these animals. Collectively our institutions spend more than a million dollars each year helping stranded animals.

Without the marine life parks, aquariums, and zoos that participate in voluntary stranding networks, who would care for stranded seals, sea otters, manatees, and seals that find themselves sick or injured on our shores? Who would continue the advances that have led to the successful rehabilitation and release of a greater number of dolphins and whales in recent years—animals that are generally very sick or injured when found?

Be assured that Alliance members take the collecting of and caring for animals very seriously. There are approximately 450 whales and dolphins maintained in all the institutions throughout North America. In 1996, 52 percent of the killer whales and 46 percent of the dolphins on display were born in these facilities.

Belugas, another commonly held species, have only been collected from Canada's Hudson's Bay, mindful that other populations have been compromised, possibly by environmental contaminants. The Hudson's Bay population is estimated by Canadian officials to be about 25,000 animals. There are reportedly 50,000 to 70,000 belugas worldwide. The average annual native subsistence harvest of these animals is 250–300 individuals *each year*. Just 34 beluga whales are housed in zoological institutions. Recent births in public display facilities bring the percentage of belugas born in aquariums to about 18 percent.

In fact, only seven whales and/or dolphins have been collected for public display in North America since 1990. On the average, that is less than one animal per year taken from the wild.

To put this in further perspective, over 8,000 cetaceans have stranded and died in the waters of the Southeast U.S. in the last 17 years. No animals have been removed from these U.S. waters in the last decade, while 4,692 animals have died there as a result of stranding.

In contrast, the percentage of animals entering public display facilities' inventories through births has skyrocketed. For all dolphins and whales, those born in zoological facilities were 8 percent in 1979, 26 percent in 1990, and 90 percent in 1995. This is one of the most striking trends in the management of marine mammals, affirming the good health care given the animals, the quality of their environments, and the success of our breeding programs.

Membership in the Alliance brings with it great responsibility for educating the public about marine mammals and their conservation, responsibility for saving stranded animals, responsibility for funding research that will help animals in our collections and in the wild. Membership also brings with it a commitment to exceptional care for the animals. That means being a responsible manager in assuring that the behavioral, medical, social, and genetic needs of the animals are met.

A responsible facility fosters thoughtful exchanges to best manage the animals in our collective care to provide for their social and behavioral needs as well as breeding opportunities. While collections of animals from the wild are minimal, that option must remain viable to sustain a healthy gene pool and to provide the opportunity to help educate the public about new species of animals. Responsible exchanges of terrestrial animals are common, understanding that it is essential to use good breeding practices to assure the health of any species. Likewise, people familiar with the raising of dogs or horses understand the merits of breeding these animals for healthy pups and foals. Inbreeding is not a responsible breeding practice as it results in physical and medical complications. Marine mammals are no different. Whales are no different. Wild populations are no different.

Thank you for taking the opportunity to learn about the importance of marine life parks, aquariums, and zoos and the public's advocacy for our many and diverse activities. We are people who care deeply about these animals. We are committed to giving them the very best care and to conserving wild dolphins and whales for your grandchildren and their grandchildren.

"*All too often, behind the facade of conservation, education, and 'family entertainment,' the grim reality of captivity consists of . . . demeaning tricks performed by increasingly traumatised . . . whales.*"

MARINE PARKS THAT KEEP CAPTIVE WHALES DO NOT PROMOTE CONSERVATION

The Whale and Dolphin Conservation Society

The Whale and Dolphin Conservation Society (WDCS) is a British charity organization dedicated to the welfare of whales and other cetaceans. The following viewpoint is excerpted from a report focusing on captive orcas (killer whales). The WDCS questions the claims made by marine parks that keeping these whales in captivity furthers education, conservation, and research. All of these goals could be better achieved, the organization argues, without capturing orcas from the wild and holding them for breeding and public display—activities that the WDCS believes lead to great hardship for the animals. The organization charges that the true purpose of marine mammal captivity is to exploit these animals for profit, which it believes does not justify the practice.

As you read, consider the following questions:

1. What distinction does the Whale and Dolphin Conservation Society make between the Sea World marine parks and the Vancouver Aquarium?
2. According to the article, what messages about ocean conservation do marine parks fail to convey?

Excerpted from "Captive Orcas: Dying to Entertain You," a report by the Whale and Dolphin Conservation Society, May 1998. Reprinted with permission.

Marine parks evolved from the same tradition as circuses, zoos and fun-fairs: in other words, to entertain the paying public and make a tidy profit for the operators. Even today, most marine parks feature a great variety of mechanical rides as well as animals performing circus-type routines. . . .

Whilst entertainment is still very much the name of the game, in recent years, attempts have been made by marine parks to redefine their purpose. Increasingly, marine parks began to feel the need to justify their existence to the wider world beyond the turnstiles. Despite the industry's attempts to present an unvaryingly rosy image, news of unacceptably high mortalities, accidents and injuries had gradually leaked out. Details of these incidents were supported in recent years by amateur videos, which undeniably showed cramped and featureless conditions, mindless circus tricks and, most disturbing of all, occasionally captured aggressive incidents or bizarre, repetitive behaviour. . . .

This viewpoint will look at the four great 'myths', namely education, conservation, captive breeding and research, put forward by many marine parks in an attempt to justify their existence. . . .

MARINE PARKS AND EDUCATION

Education: Larger marine parks, such as Sea World, have spent many millions of dollars in their enthusiasm to raise their educational credibility. In 1990 alone, Sea World claimed to have spent $3 million on its educational programmes.

On the surface, this might appear a wonderful opportunity to inform the paying public as well as visiting schoolchildren about the marine environment, using the biology and social behaviour of orcas as the focal point. By lighting the touch-paper of their enthusiasm, marine parks have the opportunity to inspire a new and profound respect for marine mammals and their ocean habitat. But how well have they responded to this challenge and what calibre of information is being disseminated through these educational programmes? Equally important, how much of the information is accurate, appropriate and in context—and how much is retained once the visitor leaves the park?

Arguably, the overwhelming obstacle to any attempt at educating visitors lies in the very fact that the captive situation bears absolutely no relation to the life of free-ranging orcas in the wild. Captive orcas are a sad caricature, a weary and spiritless version of their wild counterparts. Visitors cannot help but leave with a distorted perception of orcas and their environment. Most disturbingly, they may leave with the notion that it is acceptable to confine orcas and other animals, solely to meet human demands for entertainment and 'education'.

Attempts to inject an educational component into show routines have met with very mixed results. Marine parks steadfastly maintain that the public learns best if education is combined with entertainment. Sea World's Brad Andrews commented in 1991 that 'zoological displays are the most effective means of acquainting and educating the greatest numbers of people about wildlife. Live animals hold a person's interest in a way not possible with static exhibits.' A fair point, if the educational component was not so blatantly at odds with the entertainment value. All too often, splashy, showy routines take centre stage and an often feeble attempt at education is tagged on, almost as an afterthought. . . .

Although show scripts have improved somewhat in recent years, and are not nearly so corny, cliché-ridden and self-congratulatory as they were until the mid-1980s, the blaring rock music and stunts with the trainers nevertheless remain centre-stage at most parks. Moreover, the commentary is still full of careful euphemisms to avoid sensitive or negative connotations. Hence, captive orcas are not captured but 'collected' or 'acquired from the natural environment'; they don't live in tanks but in 'enclosures' or 'controlled environments'; they perform 'behaviours' rather than tricks. These buzzwords paint a gloss over the true nature of captivity and serve to reduce the captives to the level of performing robots. Staff and trainers are also carefully coached in answering 'tricky' questions from visitors, especially in deflecting arguments relating to captivity. . . .

Thankfully, there are exceptions. There do exist marine parks—notably the Vancouver Aquarium—which have made a determined attempt to offer an intelligent and well-presented educational programme. . . .

In 1986, Vancouver introduced a new programme entitled 'A Day in the Life', which attempted to give visitors an insight into the daily happenings of a wild orca pod. Since then, the educational component has continued to expand; 1991 saw the opening of an expanded 'killer whale habitat' and that same year, scheduled performances were discontinued, partly to reflect the new education-oriented philosophy, but partly also because their female, Bjossa, had begun to behave in an increasingly unpredictable manner. Whether Vancouver Aquarium's more enlightened approach towards the display of orcas and their more sophisticated approach towards educating their visitors will be adopted by other marine parks remains to be seen. . . .

In conclusion, it may be argued that, with the exception of

Vancouver Aquarium, educational information is all too often badly presented, inaccurate and sometimes wilfully misleading. Crucially, Vancouver Aquarium is run as a non-profit society, whilst most marine parks, such as Sea World, are blatantly run as highly profitable ventures. It seems unlikely, therefore, that education will ever seriously compete with entertainment.

CONSERVATION ISSUES

Conservation: 'Conservation' is arguably one of the buzzwords of the 1990s, but never has the need to inform and educate people about the perilous state of our planet been so urgent. Many people feel that marine parks are in a strong position to use the massive appeal of whales and dolphins as a vehicle for introducing conservation messages to the public.

In an ideal world, education and conservation would go hand in hand: marine parks would follow the lead of the World Conservation Union (IUCN) and raise awareness of the need to conserve species and ecosystems. Unless and until people are made aware of the myriad threats facing marine mammals and their habitats, they cannot possibly act to lessen or remove such threats. Only through raised awareness can action be taken to halt or reverse the damage we have already done.

Whilst Nadia Hecker of the National Aquarium in Baltimore claimed in 1991 that 'conservation is a major reason for our industry's existence' it is difficult to square this boast in the case of orcas. . . .

Orcas have never popularly been regarded as an endangered species, yet, in recent years, it has become apparent that certain orca populations are much smaller than previously estimated. Many orca researchers now suspect that whaling and large-scale capture operations in the past may have had more of an effect upon orca numbers than previously recognized, and that certain populations—previously believed to number in the thousands—may actually contain only several hundred (or fewer) individuals. For example, during the 1960s and 70s, the orca population of the Pacific Northwest was decimated by live captures, with at least 56 orcas taken into captivity and a minimum of 11 orcas killed during capture attempts.

The Southern Resident community [a closely studied orca population located off the southeastern corner of Vancouver Island] was particularly affected and, some thirty years on, is only now recovering to its pre-capture size. The transient orca pod captured off Taiji, Japan, in February 1997, was the first pod to be sighted in the area for many years, fuelling speculation that orca numbers in the Eastern Pacific may also be much lower than previously believed. Until much more is known of population parameters for orcas, the 'precautionary principle' approach demands that no further orcas should be taken from the wild. . . .

Furthermore, although orca habitats in every ocean of the world are subject to a whole host of serious threats, no attempt has yet been made by any marine park to carry out 'practical orca conservation', in the sense of protecting and enhancing the species in the wild, conserving the orcas' natural habitat and so forth. Many environmentalists believe that orcas could serve as a 'flagship species' for getting across the message that the world's oceans are in a perilous state. Not content with wreaking havoc upon the land, recent research has demonstrated that man-made toxic chemicals are widely dispersed throughout every ocean and, shockingly, no body of water, however remote from land, now remains without some form of pollution. . . .

Other threats to orcas (and other marine life) include climate change; fisheries interactions, including competition with humans for scarce fish stocks and prey depletion due to over-fishing by commercial vessels; noise pollution (from shipping, seismic testing and oil and gas exploration); coastal development; logging and other industrial concerns, and so forth. . . .

MARINE PARKS AND CONSERVATION

How have the marine parks responded to this challenge? Sadly, for the most part, marine parks have conspicuously failed to take up the challenge. Conservation messages—when they are incorporated at all—are largely tagged onto the end of a show or are buried beneath the razzmatazz. . . .

In some cases, conservation messages are cynically manipulated by marine parks in a blatant attempt to present captivity in a favourable light by comparison; thus justifying the parks' continued existence. The ocean is frequently presented in a negative light, as a dangerous place full of hazards, both natural and man-made. . . .

Opposition to captivity is presented as illogical and misdirected, given the dangers of the natural environment. 'In an age of severe ocean pollution, oil spills, tuna net deaths and death from entanglement in discarded fishing gear, certain groups have unfortunately chosen to focus their time and energy on questioning the validity of displaying marine mammals at public aquariums.' (National Aquarium, Baltimore, 1989)

Marine World Africa USA declared in 1991 that 'in one day of tuna fishing in the eastern tropical Pacific, commercial fishing fleets have killed more dolphins than have ever been collected for oceanariums.' This claim is both exaggerated and irrelevant, since it is a bit like historically justifying slavery on the basis of famine and disease in Africa! One form of injustice does not vindicate another.

Significantly, the majority of marine parks have conspicuously failed to use their high public profile and financial clout to lead the way in cleaning up the ocean environment. It has been left to environmental groups to lobby for stricter curbs on marine pollution and to raise public awareness about the millions of needless dolphin deaths during commercial tuna catches. . . .

Whilst outwardly happy to promote themselves as conservation bodies, most marine parks will only adopt the mantle of conservation as long as it equates with their overall agenda. Put simply, the parks don't often practice what they preach. . . .

CAPTIVE BREEDING

Captive Breeding: The vast majority of marine parks have a miserable record when it comes to captive breeding. All but a handful have been unable to produce a single surviving calf. Only three facilities/companies have been able to keep a captive-born infant alive for more than a matter of days. For example, Maggie, an Icelandic orca held at Kamogawa Sea World in Japan, gave birth to a male calf in March 1995. The calf died after a mere 30 minutes. Her second calf was stillborn in early autumn 1997 and Maggie herself died on October 7th, 1997. Marineland, Niagara Falls, Canada, has produced five surviving calves, one of which ('Splash') has been exported to Sea World, whilst Marineland Cote D'Azur in France has two surviving calves. In comparison, Sea World's total of 13 calves (eleven born at Sea World, two acquired from other parks) seems a creditable achievement. But is it? How successful has Sea World's captive programme really been and what are the main reasons for encouraging orcas to breed in captivity? . . .

Behind the scenes and far from the gaze of even the most inquisitive of visitors, a catalogue of miscarriages, stillbirths, calves surviving only a matter of hours or days and females dying of pregnancy-related conditions has unfolded over the years. These incidents give the lie to the 'happy family' myth propagated by the marine parks.

Why breed? The obvious, yet most cynical, answer is simply that nothing pulls the crowds like a baby animal. Sea World has more captive-born orcas than any other marine park and each new birth is heralded in a blaze of media publicity, with crowds flocking to see 'Baby Shamu'. . . .

A less palatable answer, but certainly one which carries equal weight, is that breeding must take place in order to replenish captive numbers; in other words, to replace dead orcas. Since 1985, the year of the first successful captive birth at Sea World, a total of 12 adults (eight females, four males) have died at Sea World parks alone. Ironically, Sea World has lost five adult females during pregnancy or shortly following births.

What about the much-vaunted conservation benefits of captive breeding? Captive breeding is often promoted as the great hope for conserving endangered and vulnerable species. Marine parks have certainly done their share of basking in the reflected glory of a small handful of successes in this field for example: manatees and Malaysian otters. From a conservation angle, the whole purpose of a captive breeding programme should be the eventual release of captive-bred individuals in order to replenish threatened or depleted wild populations.

This goal echoes World Conservation Union (IUCN) policy on captive breeding, which recommends that wild populations of vulnerable species should be assisted by means of a viable captive breeding programme before populations are allowed to decline too drastically. The policy specifically states that 'reintroduction to the wild should be the ultimate objective of all captive breeding programmes.'

WHALES DO NOT BELONG IN CAPTIVITY

The Humane Society of the United States opposes keeping whales and dolphins in captivity, for public display, for shows, because there's just no way that a facility can provide for these animals. Their environment is so alien to ours that in the end what you end up with is a sterile environment for them in captivity.

Naomi Rose, Frontline interview, 1997.

Captive breeding as a conservation tool is clearly not applicable in the case of orcas or bottlenose dolphins, the two cetacean species most commonly held captive. Neither species is regarded as endangered in the wild, no captive-bred orcas have been liberated and, to date, marine parks have not shown any interest in a release project. . . .

Despite Sea World's pride in its captive breeding record, it is time for the marine parks to acknowledge that, overall, the programme has been a failure. Captive births, far from swelling the captive population, at best have merely served to keep pace with deaths. Calves born merely replace dead adults. Furthermore, failed pregnancies and infant mortalities have been unacceptably high, as have maternal deaths. . . .

Marine parks have never had the intention of returning captive born orcas to the wild. Yet, if the gene pool amongst captive orcas continues to shrink, the unthinkable might happen. In an obscene juxtaposition of the ideals of captive breeding (i.e., re-populating the wild) marine parks may move to resume further captures from the wild, in order to support a shrinking and unviable gene pool in captivity. . . . If captive breeding programmes do not result in self-sustaining captive population, then surely it is time for the parks to consider whether orcas should be kept in captivity at all.

MARINE PARKS AND SCIENCE
Research:

> Most of us engage in little pure science—we function to bridge the gap between science and lay people in one very specific area. (Sea World's Otto Fad, 1994)

> No aquarium, no tank or marineland, however spacious it may be, can begin to duplicate the conditions of the sea. And no dolphin who inhabits one of those aquariums . . . can be described as a 'normal' dolphin. Therefore the conclusions drawn by observing the behaviour of such dolphins are often misleading when applied to dolphins as a whole. (Jacques Cousteau, 1975)

Marine parks have been conducting 'research' upon captive orcas and other cetaceans since the mid-1960s, when scientists working with Moby Doll, at Vancouver Aquarium, attempted to analyse orca sounds and visual capabilities. In the 30 intervening years, the larger marine parks have spent many millions of dollars in their quest to learn more about the biology and social behaviour of their orcas.

In 1991, Marine World Africa attempted to defend the confinement of orcas by asserting that 'it is vital that a handful of these animals be in captivity so that we can learn about them, physiologically and behaviourally.'

Can this really be adequate justification for keeping fifty-four orcas captive world-wide and what has really been learnt? Has research using captive orcas furthered our knowledge of the species as a whole, or merely served a useful dual role: firstly providing information applicable only to the husbandry of captives and secondly, giving a veneer of academic importance to the marine parks, elevating the status of the larger facilities from merely entertainment park to the level of a research institute?

RESEARCH IN THE WILD

In fact, by far the greatest proportion of research on orcas has been carried out in the wild. . . .

The most obvious advantage of studying orcas in their natural habitat is that 'naturalness' can be guaranteed. The research is benign (i.e., non-invasive): the orcas are simply observed going about their daily life. . . .

Researchers make use of photo-identification techniques and underwater recordings using a hydrophone, analysing the results once back in the lab. In-the-wild studies have yielded much valuable information on such diverse areas as population structure, diet, acoustics and social behaviour.

RESEARCH IN CAPTIVITY

In the highly artificial captive environment, this 'naturalness' factor disappears. Captivity, by its very nature, frequently distorts the behaviour and vitality of the orca to an unacceptable degree. Life in a wild orca pod, with the freedom to range at will, dive to great depths and associate freely within the pod is replaced by a world whose horizons are bounded by the shallow concrete pool, characterised by forced associations and by the virtual absence of freewill. Since the captive situation bears no relation whatsoever to the natural environment, much of the so-called 'research' conducted by the marine parks is only applicable at best to a tiny minority of orcas—those held captive. . . .

Whilst the larger marine parks are eager to promote themselves as research institutions, the fact is that much of the research conducted is motivated less by the desire to increase scientific knowledge of orcas than by the necessity of keeping their captives alive. Hence, the need to improve husbandry and veterinary knowledge fuels the research agenda. . . .

Surely the primary objective of research should be to expand knowledge in ways which will benefit both captive and wild populations? Research findings should be both applicable and replicable outside the research venue. Sadly, with a few honourable exceptions, much of the research undertaken at marine parks can only be described as bad science, conducted in a haphazard fashion, with much duplication of effort and often rendered meaningless because of flawed hypotheses and methodology. The overwhelming impression is that research is undertaken purely to suit the needs of the marine park and to justify the continuing confinement of orcas. . . .

The first orca was taken from the wild in 1961. In the 37 years which have elapsed since that first capture much has been learned about the species. Many people have undoubtedly gained from the experiment, not least the owners of the marine parks who have derived enormous profits from this most lucrative of assets. But the image of the killer whale peddled by the marine parks—the cuddly sea panda, who lets children sit upon its back and playfully splashes crowds with water—is every bit as misleading as the orca's previous public incarnation as a ferocious and blood-thirsty killer. Rather like an incomplete jigsaw puzzle, the true nature of the animal is hinted at, but never fully revealed. The truest portrayal of what constitutes orca society has been provided by researchers studying the species in the wild, supplemented in recent years by sensitive yet realistic coverage in natural history films and documentaries.

The species as a whole may have benefited from wider public knowledge and appreciation, but few outside the industry would argue that the captives themselves have gained from years, even decades, of incarceration. Separated from their families, kept in cramped and featureless tanks, forced to endure an unvarying routine and deprived of the opportunity to use their natural strength, speed and stamina to hunt for their own food: the captive orca is a mere shadow of its wild counterpart.

All too often, behind the facade of conservation, education and 'family entertainment', the grim reality of captivity consists of blaring piped 'muzak', demeaning tricks performed by increasingly traumatised and sickly whales, along with a growing litany of 'accidents', illnesses, stillbirths and premature deaths.

All the information presented in this viewpoint points to a single, inescapable conclusion: . . . namely, that orcas are inherently unsuited to captivity.

PERIODICAL BIBLIOGRAPHY

The following articles have been selected to supplement the diverse views presented in this chapter. Addresses are provided for periodicals not indexed in the *Readers' Guide to Periodical Literature*, the *Alternative Press Index*, the *Social Sciences Index*, or the *Index to Legal Periodicals and Books*.

Lorrayne Anthony	"The Lonely Whale," *Maclean's*, November 3, 1997.
John Carey	"Embattled Behemoths," *International Wildlife*, May 15, 1995.
Bob Chorush	"There's No Place Like Home," *Animals' Voice Magazine*, vol. 9, no. 2, 1996. Available from PO Box 16955, North Hollywood, CA 91615.
John-Thor Dahlburg	"Global Conference on Whales Debates 'To Hunt or Not to Hunt?'" *Los Angeles Times*, October 22, 1997. Available from Reprints, Times Mirror Square, Los Angeles, CA 90053.
Mark Derr	"To Whale or Not to Whale," *Atlantic Monthly*, October 1997.
Economist	"Whales: On the Menu?" October 25, 1997.
Susan Essoyan	"Group Seeks to Halt Audio Blasting of Whales by Navy," *Los Angeles Times*, February 25, 1998.
Dan Jewel	"Whale of a Tale," *People Weekly*, November 24, 1997.
Terry Johnson	"The Whaling Trade," *Canadian Geographic*, January/February 1998.
Michael Kundu	"Native Whaling in a Modern World," *Animals' Agenda*, July/August 1998.
Kenneth Miller	"Almost Home," *Life*, March 1996.
Rebecca Scheib	"The Deafening Deep: Sonar Threatens Marine Animals' Hearing," *Utne Reader*, July/August 1997.
Robert Sullivan	"Permission Granted to Kill a Whale. Now What?" *New York Times Magazine*, August 9, 1998.
Christopher Wood	"A Whale of a Debate: Captivity of Killer Whales," *Maclean's*, May 8, 1995.

For Further Discussion

Chapter 1

1. Kieran Mulvaney argues that international political efforts to protect the oceans have run into several obstacles. Do Michael Weber and Judith Gradwohl agree or disagree about the effectiveness of global agreements? On what do they base their hope for the future of the world's oceans? Explain.

2. The viewpoint by Barry Kent MacKay and Fran Stricker originally appeared in the publication of an animal rights group. Is an animal rights orientation evident in their arguments? Must one fully agree with the idea that fish have certain rights in order to accept the authors' arguments on the state of the world's fisheries? Why or why not?

3. Michael Parfit argues the extent of the fishing crisis has been sensationalized by environmental groups. Do you believe such criticisms can be applied to the viewpoint by Barry Kent MacKay and Fran Stricker? Explain.

4. After reading the viewpoint of Hans Hanson and Gunnar Lindh and the viewpoint of S. Fred Singer, who do you believe presents more convincing arguments as to whether rising sea levels pose a serious problem? How much proof do you hold to be necessary to conclude that rising sea levels warrant significant preventative action? Explain your answers.

Chapter 2

1. What new factors and considerations have made the need for international cooperation and regulation of ocean resources more important, according to Lawrence Juda? Do his arguments, in your view, neglect the role of market forces in allocating ocean resources—a factor stressed by Peter H. Pearse? Explain.

2. Ian Townsend-Gault and Hasjim Djalal argue that the Law of the Sea should be supported because the treaty helps to promote the principle that the ocean resources should be the "common heritage of mankind." Do you believe they have fundamental philosophical differences with Doug Bandow on what exactly that phrase should entail? Why or why not?

3. After reading the viewpoints by Greenpeace and Steven Nadis, which of the ten reasons given by Greenpeace to oppose nuclear waste disposal in the oceans do you consider to be the strongest? Which do you consider to be the weakest? Justify your answer.

CHAPTER 3

1. Paul MacGregor, as head of a trade group of factory trawler operators, has an obvious economic stake in whether such ships are banned. How do you think this should affect your evaluation of his arguments defending factory trawlers against criticisms by groups such as Greenpeace? Explain.

2. After reading the viewpoints of Greenpeace, Paul MacGregor, Birgir Runolfsson, and Anne Platt McGinn, do you think a system of Individual Transferable Quotas (ITQs) espoused by Runolfsson could be used to manage the North Pacific fisheries discussed in the Greenpeace and MacGregor articles? If so, would such a system work better with or without a factory trawler ban? Explain your answer.

3. In your judgment after reading the viewpoints by Bob Holmes and Christine Weber Hale, can a distinction be made between "good" aquaculture and "bad" aquaculture? If so, what are the differences between the two? Can the concerns that Holmes raises about aquaculture be addressed through reforming specific practices, or are they inherent in aquaculture itself? Explain.

CHAPTER 4

1. David Andrew Price argues that the claim that "whales are endangered" is too generalized to be meaningful because the numbers and circumstances of whale species vary greatly. Can Elaine Robbins's viewpoint be criticized on these grounds? Do her arguments apply to all whales, or just a few species? Explain.

2. Price begins his viewpoint with an account of how a Norwegian whaler's ship was damaged by environmental activists. What do you think Price was trying to accomplish with this anecdote? Do you believe it is relevant to his arguments on the status of whales? Why or why not?

3. What arguments does Greenpeace make in opposition to resumed whale hunting? Which of these claims are directly addressed in the viewpoint by William Aron? Who in your view presents the more convincing arguments? Explain your answer.

4. The Alliance of Marine Mammal Parks and Aquariums (AMMPA) argues that the American public and legislative leaders agree that the public display of marine mammals serves an important educational function. Is such popular support relevant in evaluating whether marine mammal captivity is justified, in your view? Explain.

ORGANIZATIONS TO CONTACT

The editors have compiled the following list of organizations concerned with the issues debated in this book. The descriptions are derived from materials provided by the organizations. All have publications or information available for interested readers. The list was compiled on the date of publication of the present volume; the information provided here may change. Be aware that many organizations take several weeks or longer to respond to inquiries, so allow as much time as possible.

American Cetacean Society (ACS)
PO Box 1391, San Pedro, CA 90733
(310) 548-6279 • fax: (310) 548-6950
e-mail: acs@pobox.com • website: http://www.acsonline.org

A nonprofit volunteer membership organization, ACS works to protect whales, dolphins, porpoises, and their habitats through education, conservation, and research. The society aims to educate the public about cetaceans and the problems they face in their increasingly threatened habitats. Publications include fact sheets and reports as well as the society's research journal *Whalewatcher*, published twice a year, and the quarterly newsletter *Spyhopper*.

Coast Alliance
215 Pennsylvania Ave. SE, Washington, DC 20003
(202) 546-9554 • fax: (202) 546-9609
e-mail: coast@igc.org • website: http://www.coastalliance.org

Coast Alliance is a nonprofit organization dedicated to protecting coastal resources and furthering public understanding of issues affecting the coastal areas of the nation. The alliance publishes numerous reports and a quarterly newsletter on the environment, conservation, and the effects of pollution.

Greenpeace USA
1436 U St. NW, Washington, DC 20009
(800) 326-0959 • fax: (202) 462-4507
e-mail: info@wdc.greenpeace.org
website: http://www.greenpeaceusa.org

Greenpeace opposes nuclear energy and the use of toxic chemicals and supports ocean and wildlife preservation. It uses controversial direct-action techniques and strives for media coverage of its actions in an effort to educate the public. It publishes the quarterly magazine *Greenpeace* and the books *Coastline* and *The Greenpeace Book on Antarctica*.

High North Alliance
PO Box 123, N-8390 Reine i Lofoten, Norway
(47) 76 09 24 14 • fax: (47) 76 09 24 50
e-mail: hna@highnorth.no • website: http://www.highnorth.no

The alliance is an umbrella organization that represents fishermen, whalers, and sealers from Canada, Greenland, Iceland, Norway, and various coastal communities. It is committed to the sustainable use of marine resources. Its publications include *Essays on Whales and Man* and the monthly newsletter *High North News*.

Living Oceans Program

National Audubon Society
550 South Bay Ave., Islip, NY 11751
(516) 224-3669 • fax: (516) 581-5268
e-mail: mlee@audubon.org
website: http://audubon.org/campaign/lo

The Living Oceans Program is Audubon's marine conservation program. Its mission is to improve the management of fisheries, and to restore the health of the marine environment and coastal habitats by making scientifically based analyses and recommendations to policymakers and the public. The program publishes action alerts and the quarterly *Living Oceans News*.

National Fisheries Institute

1901 N. Fort Myer Dr., Suite 700, Arlington, VA 22209
(703) 524-8881 • fax: (703) 524-4619
e-mail: www.office@nfi.org • website: http://www.nfi.org

The institute is a nonprofit trade association representing more than 1,000 companies involved in all aspects of the fish and seafood industry. The institute acts to ensure an ample, sustainable, and safe seafood supply for consumers. It offers training and merchandising publications, fact sheets, and the monthly newsletter *NFI NewsBrief*.

Seacoast Information Services, Inc.

135 Auburn Dr., Charlestown, RI 02813
(401) 364-9916 • fax: (401) 364-9757
e-mail: aquainfo@aquanet.com • website: http://www.aquanet.com

The Aquatic Network promotes sustainable use of aquatic resources and acts as a clearinghouse for information related to the ocean. It covers such areas as aquaculture, fisheries, oceanography, and ocean engineering and provides statistical databases, news articles, and publications as well as videos and software.

SeaWeb

1731 Connecticut Ave. NW, Washington, DC 20009
website: http://www.seaweb.org

SeaWeb is a multimedia educational project established by the Pew Charitable Trusts with the mission to raise awareness of the ocean and issues related to its conservation. The website lists recommended books and periodicals and publishes the monthly newsletter *Ocean Update*.

BIBLIOGRAPHY OF BOOKS

Conner Bailey et al., eds. *Aquacultural Development: Social Dimensions of an Emerging Industry.* Boulder, CO: Westview Press, 1996.

Rodney Barker *And the Waters Turned to Blood: The Ultimate Biological Threat.* New York: Simon & Schuster, 1997.

Alessandro Bonanno and Douglas Constance *Caught in the Net: The Global Tuna Industry, Environmentalism, and the State.* Lawrence: University Press of Kansas, 1996.

William J. Broad *The Universe Below: Discovering the Secrets of the Deep Sea.* New York: Simon & Schuster, 1997.

James M. Broadus and Raphael V. Vartanov, eds. *The Oceans and Environmental Security: Shared U.S. and Russian Perspectives.* Washington, DC: Island Press, 1994.

Committee on Beach Nourishment and Protection, National Research Council *Beach Nourishment and Protection.* Washington, DC: National Academy Press, 1995.

Susan G. Davis *Spectacular Nature: Corporate Culture and the Sea World Experience.* Berkeley and Los Angeles: University of California Press, 1997.

Sylvia A. Earle *Sea Change: A Message of the Oceans.* New York: G.P. Putnam's Sons, 1995.

Ross Gelbspan *The Heat Is On: The High Stakes Battle over Earth's Threatened Climate.* Reading, MA: Addison-Wesley, 1997.

John E. Heyning *Masters of the Ocean Realm: Whales, Dolphins, and Porpoises.* Seattle: University of Washington Press, 1994.

Don Hinrichsen *Coastal Waters of the World: Trends, Threats, and Strategies.* Washington, DC: Island Press, 1998.

Independent World Commission on the Oceans *The Ocean, Our Future.* New York: Cambridge University Press, 1998.

Lawrence Juda *International Law and Ocean Use Management.* New York: Routledge, 1996.

Robert W. Knecht and Biliana Cicin-Sain *The Future of U.S. Ocean Policy: Choices for a New Century.* Washington, DC: Island Press, 1998.

Peter M. Leitner *Reforming the Law of the Sea Treaty: Opportunities Missed, Precedents Set, and U.S. Sovereignty Threatened.* New York: University Press of America, 1996.

Thomas R. Loughlin, ed. *Marine Mammals and the Exxon Valdez.* San Diego: Academic Press, 1994.

Joseph MacInnis, ed. *Saving the Oceans.* Buffalo, NY: Firefly Books, 1996.

William McCloskey   *Their Fathers' Work: Casting Nets with the World's Fishermen.* New York: McGraw-Hill, 1998.

Thomas Gale Moore   *Climate of Fear: Why We Shouldn't Worry About Global Warming.* Washington, DC: Cato Institute, 1998.

David B. Morris   *Earth Warrior: Overboard with Paul Watson and the Sea Shepherd Conservation Society.* Golden, CO: Fulcrum, 1995.

Roger Payne   *Among Whales.* New York: Scribner, 1995.

Peter B. Payoyo   *Cries of the Sea: World Inequality, Sustainable Development, and the Common Heritage of Humanity.* Boston: Martinus Nijhoff, 1997.

S. George Philander   *Is the Temperature Rising? The Uncertain Science of Global Warming.* Princeton, NJ: Princeton University Press, 1998.

Carl Safina   *Song for the Blue Ocean: Encounters Along the World's Coasts and Beneath the Seas.* New York: Henry Holt, 1998.

April Pulley Sayre   *Exploring Earth's Biomes: Ocean.* New York: Twenty First Century Books, 1996.

Philip M. Scanlon   *The Dolphins Are Back: A Successful Quality Model for Healing the Environment.* Portland, OR: Productivity Press, 1998.

Carl J. Sindermann   *Ocean Pollution: Effects on Living Resources and Humans.* Boca Raton, FL: CRC Press, 1996.

Peter J. Stoett   *The International Politics of Whaling.* Vancouver: University of British Columbia Press, 1997.

Jon M. Van Dyke et al., eds.   *Freedom for the Seas in the 21st Century: Ocean Governance and Environmental Harmony.* Washington, DC: Island Press, 1993.

Hillary Viders   *Marine Conservation for the 21st Century.* Flagstaff, AZ: Best, 1995.

Paul Watson   *Ocean Warrior: My Battle to End the Illegal Slaughter on the High Seas.* Toronto, Ontario: Key Porter Books, 1995.

Michael Weber and Judith Gradwohl   *The Wealth of Oceans.* New York: W.W. Norton, 1995.

Peter Weber   *Abandoned Seas: Reversing the Decline of the Oceans.* Washington, DC: Worldwatch Institute, 1993.

INDEX

protection for, 39, 66–67
Marine Mammal Commission, 175
Marine Mammal Protection Act, 114,
    177
marine mammals, 19
    captive breeding of, 189–91
    captive vs. wild, 185–86
    conservation in marine parks for,
        178–79
    opposition to, 188–89
    education/research on, 177–78,
        179–81
    efforts to protect, 39
    rescue/rehabilitation of, 181–82
    in marine parks vs. wild, 192–93
    see also whales
marine parks
    captive breeding in, 189–91
    conservation in, 178–79
    opposition to, 188–89
    education/research in, 179–81
    opposition to, 185–87, 191–92
    responsibilities of, 182–83
    support for, 177–78
Marine Research Institute (MRI), 131,
    132
Marine Stewardship Council, 25, 140
Marine World Africa USA, 188–89, 191
Markels, Michael, Jr., 76
Marx, Wesley, 40
McGinn, Anne Platt, 133
media, 30
    dramatization in, 47
    on overfishing, 41
Mexico, 148
migration, exotic, 21–22
Milazzo, Matteo, 137
Miles, Edward, 77, 104
Milleman, Beth, 20–21
Ministry of Fisheries (New Zealand),
    131
mollusks, 144, 145, 146
Monterey Fish company, 148
Mulvaney, Kieran, 17
mussels, 143, 144
    safety of farmed, 146–47

Nadis, Steven, 98
Namibia, 139
Nash, J. Madeleine, 114
National Academy of Sciences, 20
National Fishworkers' Forum, 25
National Marine Fisheries Service
    (NMFS), 113, 118
National Research Council, 19, 20, 114
National Shellfish Sanitation Program

(NSSP), 146
natural resources
    conflict over, 67–68
    exploitation of, 64–66
    and population growth, 31
Netherlands, 130
    quota system in, 136
    and seabed burial, 103–104
nets, 41
    see also trawlers
Newfoundland, 40
New International Economic Order,
    86–87
New Zealand
    fishing monitoring in, 139
    property rights management in,
        76–77, 130–31
    quota system in, 136, 137
Niagara Falls, 189
1996 Protocol to the Convention, 96–97
North Pacific
    bycatch in, 110, 121
    overcapitalization vs. catch limits in,
        111–12, 119
    small vs. large vessel fleets in, 118
North Pacific Fishery Management
    Council, 112
North Sea, 19, 20
Northwest Atlantic groundfish, 19
Norway, 42
    property rights management in, 130
    subsidies in, 137
    whale population in, 169
    whaling in, 24, 160–61, 164–65,
        166, 175
Norwegian Sea, 103
nuclear waste
    dumping of
        environmental reasons against,
            94–95
        international reasons against, 96–97
        international law on, 103–104
    see also seabed, burial

Ocean Dumping Act of 1972, 95
Ocean Dumping Ban Act of 1988, 95
oceans
    are healthy, 46
    conservation
        consumer movement for, 25–26
        hope for, 28, 33–35
        is unrealistic, 23–25, 26
        international, 23–25
        key trends for, 31–33
        role of humans in, 30–31
        whale conservation and, 162